SEVEN PAPERS

A collection of investigative papers on the creation of the modern brain

Dan M. Mrejeru

Copyright © 2021 A Terrestrial's Mind Publishing

All rights reserved. No part of this publication may be reproduced, distributed, or transmitted in any form or by any means, including photocopying, recording, or other electronic or mechanical methods, without the prior written permission of the publisher, except in the case of brief quotations embodied in critical reviews and certain other noncommercial uses permitted by copyright law. For permission requests, write to the publisher, addressed "Attention: Book Rights and Permission," at the address below.

Published in the United States of America

ISBN 978-1-953904-97-3 (SC)
ISBN 978-1-955243-20-9 (Ebook)

A Terrestrial's Mind Publishing
222 West 6th Street
Suite 400, San Pedro, CA, 90731
www.stellarliterary.com

Order Information and Rights Permission:

Quantity sales. Special discounts might be available on quantity purchases by corporations, associations, and others. For details, contact the publisher at the address above.

For Book Rights Adaptation and other Rights Permission. Call us at toll-free 1-888-945-8513 or send us an email at admin@stellarliterary.com.

CONTENTS

First Paper
A prehistoric C14 isotope and the wave of extinction — 4
A high prehistoric concentration of atmospheric C14 isotope — 5

A path to extinction — 18

Why prehistoric people migrated out of Africa? — 28

Second Paper

Absorption of C14 isotope stimulates neurogenesis — 36
The role played by biological absorption of C 14 in stimulating

neurogenesis and biophoton production — 37

Third Paper

A language - ready brain — 67
Changes in brain anatomy inflicted the development of language — 68

Fourth Paper

Changes in brain lateralization — 89

Conjugates switched brain lateralization and generated

a new intelligence — 90

Fifth Paper

Universal technology of language and intelligence

of complexification — 100

Universal technology of language — 101

Language brought us the intelligence of complexification — 136

The loss of language may cause the civilization to collapse — 144

Sixth Paper

Homo loquens   **150**

The hypothesis of Homo loquens   151

Seventh Paper

Challenges   **182**

Neural challenges of the gaming brain   183

A chronic metabolic oxygen deficiency   204

My motivation   220

First Paper
A prehistoric C14 isotope and the wave of extinction

Dan M. Mrejeru

A high prehistoric concentration of atmospheric C14 isotope

The Reviews of Geophysics, Volume 57, Issue 3, had published on May 29, 2019, an article by J. E. T. Channell and L. Vigliotti under the title *"The Role of Geomagnetic Field Intensity in Late Quaternary Evolution of Humans and Large Mammals."*

I will quote from its *Plain Language Summary* placed at the beginning of this paper.

"The strength of the Earth's magnetic field in the past, recorded by rocks and sediments, provides a proxy for a past flux of ultraviolet radiation (UVR) to Earth's surface due to the role of the field in modulating stratigraphic ozone. About 40,000 years ago, mammalian fossils in Australia and Eurasia record a significant die-off of large mammals that included Neanderthals in Europe. In the Americas and Europe, large mammalian die-offs can be linked to minima in Earth's magnetic field strength, implying that UVR flux variations to Earth's surface influenced mammalian evolution. For the last 200,000 years, estimates of the timing of branching episodes in the human evolutionary tree, from modern and fossil DNA and Y chromosomes, can be linked to minima in field strength, which implies a long-term role for UVR in human evolution. New and fossil find improved fossil dating, knowledge of the past strength of Earth's magnetic field, and refinements in the human evolution tree are sharpening the focus on a possible link between UVR arriving at the Earth's surface, magnetic field strength, and events in mammalian evolution".

The paper insists that during geomagnetic polarity reversal, also occurs a very low field intensity, and these factors combined have been the probable cause of large extinctions on Earth's biosystems.

I would not enter the authors' details on the biological mechanism affected by the UVR penetration on the ground.

However, I found it significant that such geophysical/biological discussions appear more frequently in the literature.

My paper's scope is to discuss the role of the high concentrations of the C14 isotope on the evolution of the human brain and the approximative timing of such occurrences.

* * *

NTD Resource Center made an informative note that I will quote here. "Cosmic rays enter the earth's atmosphere in large numbers every day, and when one collides with an atom in the atmosphere, it can create a secondary cosmic ray in the form of an energetic neutron. When these energetic neutrons collide with a nitrogen-14 atom, it turns into a carbon-14 atom and a hydrogen atom. Since Nitrogen gas makes up about 78 percent of the Earth's air, a considerable amount of Carbon-14 is produced by volume. Carbon-14 atoms combine with oxygen to form carbon dioxide, which plants absorb naturally and incorporate into plant fibers by photosynthesis. Animals and people take in carbon-14 by eating the plants. Maybe one in a trillion carbon atoms is carbon-14.

Both Carbon-12 and Carbon-13 are stable, but Carbon-14 decays by very weak beta decays to Nitrogen-14 with a half-life of approximately 5,730 years. After the organism dies, it stops taking in new carbon.

Examples of the type of material that radiocarbon can determine are the ages of the wood, charcoal, marine, fresh shells, bone and antler, peat, and organic-bearing sediments. It also accumulates in carbonate deposits such as calcite, dissolved carbon dioxide, carbonates in ocean, lake, and groundwater sources, in the caves' stalactites.

The recent summarization of geophysical studies published online by Cambridge University Press on 18 July 2016 indicates that new measurements for the paleomagnetic field strength show that the period before 12,000 years ago is characterized by low dipole moments or low strength. High values are associated with the Lake Mungo excursion between 32,000 and 28,000 years ago. In opposition to this, a field maximum has occurred 10,000 years ago and again 3,500 years ago.

* * *

Recent studies (2014) on a significant increase in the atmospheric concentration of C14 isotope during atmospheric testing of the atomic bombs found a critical effect on Dentate Gyrus's mechanism in the hippocampus.

The atmospheric tests of the atomic bombs, during the nuclear explosion, generated gamma-rays, which reacted with atmospheric Nitrogen-14 producing Carbon-14 isotopes.

The analysis and the experimental data of the mentioned effect suggest that, during that atmospheric testing era, neurogenesis's capacity almost doubled due to doubling in the atmospheric C14 isotope. There is known that during the geomagnetic excursions, the planetary shielding against cosmogenic radiation also diminishes in the range of 30-60%, allowing an accumulation of atmospheric C14 isotope 30-60% higher than usual.

However, the exact evolution of each such excursion's mechanism and the effect is not very well known, but this knowledge could be significantly improved. Each excursion's radiative variation can be better disclosed by a more comprehensive analysis of the geophysical and geological data at distinct planetary locations while employing novel correlations within other sources for C-14 data (tree rings, stalactites, sediments, pollen, corals, etc.). Maps representing the relative path of the excursion could be drawn more precisely. Graphs can suggest a variation of the field intensity along the length of the excursion at various locations. Because of the influence of locality during the excursions, the interpretation of maps and graphs would allow the specialists to establish quantitative values of geomagnetic field intensity at distinct locations during particular time intervals defined along with the duration of each excursion.

In the end, each excursion would be represented by a map that contains graphs of each locality. The charts would indicate the exact concentration of C-14 at a location and how it varied during the excursion.

Ultimately, the maps would indicate the variation of atmospheric C14 isotope at each locality along with the entire excursion interval. Undoubtedly, such a project implies tremendous work that would extend over many years.

However, the result would show what exact pulses of increased neurogenesis have occurred in each region and during which period.

Such a study could collaborate well with archaeological information available for each locality or an extended area.

Here would be produced many maps, and the specialists would interpret the possible superposition.

The final result would indicate how different populations were or were not subject to such a radiative effect. It is essential to know such information because of the considerable population mobility over the Eurasian landscapes.

We have today several planetary zones where the inhabitants display prehistoric types of culture. We call them "primitives," but it is essential to know if such "primitivism" may represent only a "blind spot" on the radiative map. If this would be the case, then it means that we all have been significantly more or less affected by the radiative component caused by the combination of the geomagnetic excursions and the cosmogenic factor.

It will change our ideas about human and biological evolution on this planet.

Probably, it would change the way we see and understand the universe as a whole and our relationship with it.

In the last 40,000 years, the geomagnetic excursions occurred quite frequently, while they affected the northern hemisphere's landmass, especially in the areas placed north of the 40 degrees north parallel. However, this represents the area where live and evolved the bulk of the planetary population and where the civilization evolved more rapidly than in other regions of our planet.

Hence, the suggestion made in this paper may help the specialists better understand what exactly shaped the evolution of human groups at various localities, and eventually, why the people have migrated so widely?

* * *

I like to present a short record and some data analysis available for the past 50,000 years on the atmospheric accumulations of the C14 isotope.

Most studies accept that C14 isotope concentration started to rise around 50,000 years ago and increased continually until 35,000 years ago. After that peak, the C14 concentration took a diminishing path until nowadays.

The graphs used by these studies indicate that during the mentioned peak, the produced value of C14 was about 600-800 per million that represents an increase of 60-80%, and which is comparable with the rise of C14 recorded during the atomic bomb atmospheric testing that was 80-100%.

The atmospheric concentration of C14 was about 300 per million about 11,000 years ago (or 30% higher than current values). It was 500 to 600 per million in the era 17,000 to 29,000 years ago (or 50-60% higher than present values).

It means that from 42,000 years ago, it had increased almost to double (or 100%) until 35,000 years ago, and it continued to remain high the entire era until 11,000 years ago (during the Younger Dryer cold epoch) when it still was 20-30% higher than current values. The field intensity suddenly increased only 10,000 years ago and for a short time.

After that increase, the field intensity dropped again for almost 5,000 years or until 4,000 years ago, but then, a new increase developed and reached its peak 3,500 years ago. In the last 2,000 years, the field intensity was constant.

However, <u>this image of a higher concentration of atmospheric C14 for almost 30,000-34,000 years must imply a significant re-analysis of the modern brain's evolution</u>. This epoch stays in contrast with the previous era between 70,000-50,000 years ago when the C-14 concentration was low or average.

However, the C-14 concentration was high again between 80,000-70,000 years ago that was the era of the first art expression at the southern end of Africa.

I would say that these 30,000-34,000 years represented precisely the epoch when the prehistoric human brain changed into that modern brain that we have today.

Nevertheless, based on other estimative sources, after the Lake Mungo phase ended (32,000-28,000 years ago), the excursion path continued with Mono Lake (28-26,000 years ago), Gothenburg (13-12,000 years ago), Solovki (7,000-4,500 years ago), and Sterno-Etrussia (2,800-2,600 years ago).

I like to specify that each excursion's time definition varies from one source to another. It is based on local geological and other information that is distinct from one locality to another.

Dipole intensity (field intensity) reached a minimum 5,500 years ago, during Solovki excursion, but rapidly increased back, reaching a maximum of 3,000-3,500 years ago. It dropped 30-50% again during the Sterno-Etrussia event (2,800-2,600 years ago), which lasted only 200 years.

It is assumed that a much more recent excursion occurred in the interval 774-775 AD.

The geophysics literature reports an extraordinary C14 significant increase (20%) from 5481 BC to 5471 BC that is an event that occurred just 400-500 years before the starting of the Solovki event, and it was associated with a solar minima phenomenon.

All the above analysis of past C14 high atmospheric concentrations allows one to assume that, besides the bulk of 60-80% increase, several other increases indicate 20-50% above normal C14 levels. It could be assumed that a case defining higher neurogenesis and increased plasticity in prehistoric brains developed in such epochs, which represented a unique radiative contribution for the entire interval of the last 250,000 years of human evolution. It may suggest that the same processes contributed to genetic changes, as in the initiation of new haplogroups.

It may stimulate the very migration, too. It is possible that a deep investigation into such processes would unveil a mechanism that promoted mass human migrations beyond the known cause of climate change.

Other studies propose that the <u>geomagnetic field's magnitude and direction modulate cryptochrome (CRY) activity</u> by influencing photochemical radical pair intermediates within the protein. It shows that exposure to a magnetic field (100 mT) is sufficient to potentiate light-activated cryptochrome's ability to increase neuronal action potential firing.

It proves that cryptochrome's activity is sensitive to an external magnetic field, like the geomagnetic field variation, capable of modifying animal behavior and causing gene expression changes.

Part two

I like to mention and quote two articles that I found only later and provide precise data about the C14 isotope concentration from 54,000 years ago to Classical Antiquity (8-3 Centuries BCE).

The first paper I like to mention here is "*Analysis of the atmospheric C14 record spanning the past 50,000 years derived from high-precision Th230, U234, U238, Pa 231, U235 and C14 dates on fossil corals*", written by Tzu-Chien, Richard G. Fairbanks, Li Cao, and Richard A. Mortlock, published in Elsevier (Quaternary Science Review 26) on June 19, 2006.

The study presents several graphs that I like to discuss in this short addendum to my original paper. In my previous article, I was forced to make several assumptions based on the literature's lesser information.

The first graph I like to discuss displays radiocarbon production, tree-rings, Kiritimati, Barbados, Araki corals, and Araki and Barbados corals.

Here the graph indicates an increase in *carbon production* starting 48,000 years ago that reaches a peak around 41,000 years ago with a 60%-80% increase over typical values. It drops to 23-25% 36,000 years ago, to increase again about 34,000 years ago to 40%. It stays between 10% (33,000 years ago) and 32-33% since 22,000 years ago when it reached a new high of 38% around 22,000 years ago. It remains between 35% and 10% until 8,000 years ago, when during Solovki excursion (7,000-4,500 BCE), it records another 28%, and another one during Sterno-Etrussia excursion (around 2,700-2,400 BCE) at 15%. In the graph, the *tree rings* follow exactly radiocarbon production.

What is discrepant here, there were the values given by the corals. The corals show peaks significantly higher: for the epoch 42,000-37,000 years ago, and the values were 60 to 70%. And until 22,000 years ago, the values remained above 50%. From 22,000 to 18,000, the values stayed at 45%. Between 16,000 to 10,000, they were 45-48%, and 6,000 years ago, they reach values of 30%.

The other graphs of this paper show similar results.

The authors comment that the high C14 concentration decays slowly, while the increase is rapid when another excursion strikes shortly after that.

In sum, this paper indicates that the highest peak (almost 70-80%) is reached during the Laschamp excursion (42,000 years ago) and stays elevated for the next 3,000 years.

The second paper I like to mention is *"Atmospheric C14/C12 changes during the last glacial period from Hulu Cave"* (China), written by Hai Cheng, R. Lawrence Edwards, John Southon, Katsumi Matsumoto, and a long list of colleges.

They have defined a more precise age measurement on several stalagmites in Hulu Cave (China).

They concluded the following:

-54,000 to 50,000 years ago, the value was **12%** increased;

-from 50,000 to 43,000 years ago, the value increased another 14%, reaching a total of **28%**;

-the Laschamp excursion presented a short-term (temporary) reversal, and the reversal effects made this excursion significantly stronger than the following geomagnetic events, which randomly continued to occur for the next 30,000 years;

-from 43,000 to 38,000 years ago, the values reach at its peak over **70%**, and on average it was **60%**, due to the high strength of Laschamp excursion;

-from 38,000 to 25,000 the values were **60% to 40%**;

-20,000 years ago, the average value was **50%**.

In this first interval of 23,000 years long, the values were increased 50-60%.

The last interval occurred from 10,000 to 2,500 years ago:

-11,000 years ago, the value was 15%;

-the value increased again 6,000 years ago (30%) and remained high for more than 1,500-2,000 years;

-the last increase occurred 2,500 years ago.

However, all this information confirms my statement from the original paper that for at least 30,000 years, our ancestors lived within a high atmospheric C14 concentration that was in the range of 50% to 30% higher than the average recorded before 1950.

There is known that the activation of a wide number of differentiated signaling pathways for a population of neurons requires high entropy.

Such activation defines a pluripotency state, where cellular diversity dominates, and it represents an undifferentiated/diversified state.

At a single-cell level, the high network entropy is thought of as a signaling pathway measure. High-entropy would maintain open most options of the cell, which are associated with diverse cell fates. Such an undifferentiated state defines the network "plasticity."

Energy flow is fundamental for hippocampal dynamics, including the neurogenesis processes developed in the Dentate Gyrus placed in the hippocampus. Cross-frequency coupling is a manifestation of a nonlinear energy transfer.

Here, the energy of C14 isotopes is transferred to the neural network, increasing the entropy of this system of cells locally. This transfer gives a bust to the neurogenesis process in Dentate Gyrus and affects the newly generated neurons' fate and options.

The research found out that hippocampal rhythm underpins language processing. This could explain why the high concentration of C14 isotopes in that the atmosphere in the era mentioned above has stimulated human brain plasticity, contributing to environmental adaptation by higher levels of "plasticity" and the language development as another form of "plastic" adaptation.

Language is an undifferentiated complex, where the "plasticity/flexibility" of the neural network helps adapt to various options, like communication, planning, control, and socialization. Higher "plasticity" offers more "options" that are capable of dealing with higher "environmental uncertainties."

All these mentioned elements contributed to Out-of-Africa Migration's success, where the language was the core technology employed to investigate and adapt.

There is no coincidence that those humans went out-of-Africa cca. fifty thousand years ago, get a big help cca. forty-three thousand years ago, to adapt to new environments by intermediating an increased atmospheric concentration of C14 isotope. C14 isotopes increased the entropy locally.

Here, there is to say that the entire migration may have failed to succeed without such a lucky occurrence because there initially was not sufficient support to adapt to new environments.

There is known that an increase in neuronal network entropy increases oxidative stress and free radicals. De-excitation of free radicals produces biophotons. Short-pulses of augmented entropy generate biophotons and imply short-phases of biophotonic entanglement. There is to say that such entanglement was the motor of correlative thinking, discoveries, innovation, and creativity, which all contributed to general human adaptation.

As a result of the de-excitation process, biophotons' emission increases neural network communication, improving cognition and memory storage, which further contributes to developing the language.

As it appears, during the mentioned era of low intensity of the geomagnetic field, pulses of gamma rays penetrated to the ground of our planet. Consequently, they generated pulses of C14 isotopes.

A path to extinction
Who survived 42,000 years ago, and how?

Recently has been an intense scientific debate in the professional media on the biophysical consequences on the biota of the Laschamp Geomagnetic Excursion of 42,000 years ago. The main issue was how an alleged high UV radiation level correlated to a documented extinction of megafauna.

However, my previous focus in other papers was on an extinction affecting 42,000 years ago the hominins (Neanderthals, Denisovans, and all others documented or not) and Homo sapiens proper.

In the view of several scientists, the Laschamp event has produced a higher than normal level of Ultraviolet (UV) radiation that severely endangered the megafauna, causing its extinction. Allegedly, such megafauna was unable to reach for cover against the UV radiation, suffering lethal consequences.

As documented by the said group of scientists, the UV radiation was detected in samples dated for the time of this geomagnetic occurrence. For example, it was three times higher near the Antarctic continent's shores.

Laschamp excursion was documented as a temporal reversal of magnetic polarity that lasted for most of its excursional duration of 400-1,000 years.

We know that UV radiation can reach higher levels that may be dangerous at the ground level when it penetrates through ozone holes formed in the terrestrial stratosphere and atmosphere.

The scientific literature describes how ozone holes are generated naturally every year from October to the early spring near the Antarctic continent.

Geophysical literature indicates that a significant previous geomagnetic event, call the Blake Excursion, took place from 131,000 to 119,000 years ago and encountered an abrupt but complete reversal of polarity that hovered for 6,000 years.

A more recent international study, analyzing specific markers in a cave of South China, came with additional polarity reversal information; accurately was determined a group of other two reversals, where the first occurred for several centuries between 106,000 and 103,000 years ago, and the second was an abrupt reversal between 98,000 and 96,000 years ago.

Hence, there were three polarity reversals in the interval 131,000 to 96,000 years ago, where the main one lasted for 6,000 years.

Coincidently, the archaeologists defined and dated the skeletal appearance of "modern" Homo sapiens around 130,000 years ago, marking a sharp distinction compared to an early archaic type of Homo sapiens dated 190,000 years ago. Within the same coincidence range, one can relate to another long geomagnetic excursion, Biwa I (182,000-171,000 years ago), with its two documented polarity reversals.

However, another coincidence refers to the Norwegian-Greenland Geomagnetic Excursion (80,000-72,000 years ago) that overlaps the first expression of art manifested in the ocean shore caves of the southern peak of South Africa. But, maybe more important was an all-time geomagnetic minimum intensity recorded around 65,000 years ago, and which overlap an alleged out-of-Africa migration-tentative that deadened one in South India and one in South China.

A more massive and successful migration can be documented 55,000 years ago, which seems to reflect another mental and biological change in Homo sapiens during the all-time minimum intensity event of 65,000 years ago.

I mentioned all these geophysical events to indicate that the Laschamp event was not unique in any geophysical way. But it became a starting point for a chain of other geomagnetic sharp intensity variations, where multiple excursions or semi-excursions have been partially or fully documented for the interval 37,000 years ago to 2,500 years ago.

I mentioned all such geophysical events because they allow the cosmogenic radiation to penetrate to the ground surface, affecting biota.

Suppose the penetrating UV radiation was the case in all six similar reversal situations. In that case, we should have as many animal extinctions as the number of these excursions, temporal reversals, and minimal intensity episodes indicate. They may have caused similar amounts of dangerous ozone holes all over the planet like they allegedly occurred during the 42,000 years old event. Within such waves of extinction, the Homo genus and the majority of other animals prevailed and survived.

As a fact, it seems that the megafauna survived well the first five waves of alleged extinction, but it couldn't make it through in the sixth one.

I do not try to be ironic, but here, I consider that some common sense must guide our scientific investigative spirit.

We have to relate the tool offered by interdisciplinarity to the geophysical incidences, the working of the biological mechanisms, other non-apparent causes, and determine what determines what, why, and how?

In my research, I find debatable information that tries to explain ozone holes' production within a geophysical event, like the one that implies temporal polarity reversal. This explanation I found suffices only for The Antarctic continent neighborhood, and it does not apply well to other terrestrial areas. I found some additional scientific information to relate the cyclical formation of the Antarctic ozone hole to the mechanism that produces the southern magnetic anomaly.

Now, I will move the discussion to the issue of the cover taken by humans and humanoids in the caves.

I like to remind the reader that around 42,000 years ago, the global climate experienced significant warming.

Because the archaeologists found various skeletal remains in the caves, one could assume that the cavemen were hunter/gatherers who spend most of the daylight outside the cave to gather those products necessary for their feeding. Also, they bring to the shelter of the cave such products, eventually storing them.

Hence, the hunter/gatherers mostly spent their daylight outside of the cave.

Another aspect is that 42,000 years ago, the Homo sapiens were well out-of-Africa, rapidly advancing into Europe from the east to the west; other sapiens were moving from Central and South Asia to the east, west, and north of the continent; while still other sapiens already had a good foot in the New Guinea and Australia.

As the process of migration implies, the participants continually moved with little appetite to waste time finding the shelter of hidden caves. Hence, migrants mostly traveled crossing through open spaces, where no particular radiative cover was immediately available.

It is known that caves predominantly develop in the karstic (limestone) environments. Not all geographic regions abound in such karsts. Most of the karst caves are crossed by the running water flows, which cause sudden flooding. Also, the cave would naturally be populated by many species, which would not readily accept human cohabitation.

I like to quote Margaret Conkey, Professor Emerita at the University of California, Berkeley. I will mention her interview given to Nautilus on December 5, 2013. She initiated for 20 years a study named "**Between the caves**." She concluded that Paleolithic people were much more than cavemen.

She said in the interview: "*Caves are constrained spatially, preservation is excellent because they're usually limestone and alkaline, which helps preserve bone and other materials that don't often preserve in the open air. But caves are an unrepresentative sample of where people were and what they did. People weren't hunting in the cave, they weren't collecting raw materials in a cave, they weren't collecting firewood or other things. So where were they the rest of the time, and what were they doing?*"

"Almost all caves are described by archaeologists as seasonal, namely as autumn or winter occupation."

"*We found many Paleolithic open-air sites, but we can't determine exactly what period because we just don't have any datable, organic materials. The number of artifacts we found suggests a long-time use of the landscape-people coming to these areas probably 80,000 years ago and even into Neolithic.*"

"*My colleagues and I are suggesting that we have certain biases about what constitutes a "home" and what mobile people didn't think of home as a stationary physical structure. Clearly, based on what we found,* **our ancestors were way more spatially ambitious than the caveman we thought them to be**."

Recently, archaeological literature begins to abound with information about the discovery of open-air Paleolithic sites.

The sites run from Jordan Valley, Sinai, and the Levant to all over Eurasia, Central and North Steppes, deserts, and forests. In the open-air sites were found Homo sapiens skeletons, but also Neanderthal and Denisovans remain. Hence, **all hominins shared the culture of extended open-air living**.

It is known that the geomagnetic field becomes significantly weaker during the geomagnetic excursions, allowing cosmogenic and galactic radiation to penetrate the terrestrial grounds.

Such radiation reacts with the atmospheric components, producing nuclide showers, where the C 14 isotopes and B 10 are prevalent. When the biota absorbs C 14, each organism reacts in a distinct metabolic manner, but overall, this stimulus generates a metabolic production of reactive oxygen species (ROS). Also, the Ultraviolet (UV) radiation induces an additional metabolic production of ROS.

The dangerous UV radiation level seems to reach the ground only under the weakened cover of ozone holes placed in the stratosphere and atmosphere.

The concern about ROS is related to its stimulation of free-radical production. The scientific literature describes how the biological system's dangerous levels are eradicated within an immediate process of deexcitation, where the formed free-radicals are transformed into biophotons.

In general, the malfunctioning of deexcitation allows the free-radicals to evolve into cancers. Another bionegative effect of ROS concerns the eyes, where it favors the building of cataracts. Still, the most common result of UV radiation is skin cancer.

Regarding the potential harm of UV exposure on various animals, it is a view that small animals could easily find the cover of the shadowed areas. In contrast, the large ones could not benefit from such an advantage.

How would such small animals and everyone else sense the need to seek cover against the sunlight? How would animals' lifestyles and behaviors, as they must continually search for cover, be affected during many generations of such phenomenon's exposure?

As it is known, the impact of the UV radiation, when the ozone holes are not present, is biologically smaller than the effect caused by absorbed C 14 isotopes.

Hence, during the Laschamp event, some additional metabolic ROS production was prevalent in all biota. Still, it was proportional to each organism's natural protection, like skin, fur, habitat, geographical elements like altitude, etc.

It is known that large animals have a 30-40% reduction in their metabolism than the rest of the animals. It has been documented how the metabolism's speed relates to a species' life duration. The metabolism of large animals is associated with their long-life duration.

A study found that most megafauna's reduction of metabolic speed is less than half than for small animals.

An additional metabolic ROS production in megafauna would accelerate their metabolic seed to a direct proportion to the increased amount of C 14 in the atmosphere or the elevated UV radiation level. Such an acceleration was shortening their life duration by almost half. But this diminishing of the lifespan would not change their reproductive rate.

The reproductive rate is setup naturally, and accidental causes cannot change it. It is a negative relationship or an inverse proportionality between reproduction and lifespan. It means that animals that live longer have a proportional lower reproductive rate.

Consequently, it results in an imbalance leading to gradual extinction in a few generations. It may gradually diminish individuals' number within several hundred years until the entire species would perish.

Hence, the megafauna did not disappear instantly, but along the entire interval of 1,000 years of more, when the atmospheric concentration of C 14 isotopes and the UV radiation levels were at their peak.

Now, I like to discuss another aspect that is the effect of ROS (from C 14 and UV) on animal neurogenesis. It is known that ROS's metabolic higher levels stimulate the production of neurons destined for the olfactory bulb function and some other sense replacement needs. Such neurons have in most animals a concise life, like six weeks for mice.

Additional metabolic ROS will supplement the required replacement to their olfactory function and others' senses, making these animals better disseminate in their environment.

As the research has demonstrated, human neurogenesis is highly driven to cognition functions while minimally toward olfaction and other senses.

Additionally, in humans, the cortical regions, supplemented by neurogenesis and subserving the high cognitive functions, have a prolonged development path beyond birth that extends from mid-and-late adolescence into adulthood. The exact process is much shorted in other animals.

As in other papers, I wrote in detail, the Homo sapiens did not properly survive because it acquired a new mental setup that caused a language-ready brain and other plastic features to extend to the biological level. Homo sapiens disappeared when its twin transformation replaced it. I called this twin Homo loquens.

Thus, not only the Neanderthals and Denisovans vanished but all other hominins along with Homo sapiens proper.

At the beginning of this paper, I presented a quick view of the geophysical events of the last 200,000 years. Now, one can assume that archaic Homo sapiens appeared during the Biwa I event (190,000 years ago) and evolved, along with other hominins, under the influence of the other occurring geomagnetic events, especially during the Blake event (131,000-119,000 years ago) that established another marker in Homo sapiens, producing its "modernity."

Somewhere on this route, Homo sapiens and the Neanderthals encountered a neurogenetic change that prompted the cognition to the detriment of other senses. In the case of the Neanderthals, it is documented that they experienced an active cultural life, and just before their demise, a significant artistic and cultural peak was achieved. Thus, their neurogenesis favored cognition almost similarly like in Homo sapiens.

Even then, an essential neurological variation occurred and drove them apart. One can assume that the Neanderthal and other hominins neural circuitry and other neural connections were selected and evolved differently from Homo sapiens.

Neurogenesis in sapiens helped transform their brain toward language production and saved them from extinction by transforming the sapiens into a mentally distinct species.

I hypothesize that an adverse path of neurogenesis manifested in Neanderthals, and most probably, it also occurred in their Denisovan cousins and other hominins. When 42,000 years ago, well-stimulated neurogenesis introduced avalanches of new neurons. These hominins did not have the proper circuitry and neural connections, and the effect was undesirable and harmful because it generated brain disorders acute mental diseases. In few generations, the catastrophic hominin change in behavior drove them extinct.

At present, some 26-30% of our adult population has a diagnosable mental disorder. Suppose an unexpected situation of an increase by 30-50% in the current proportion may occur. In that case, one may estimate that the human species, ultimately plagued by acute mental diseases, would reach the edge of extinction in a few generations.

Medical literature indicates that human' mental diseases shorten the lifespan by 10-20%. But here, one should consider that today humans benefit from significant medical and life support assistance that alleviates the disease's burden and effects, significantly extending such patients' lives.

The hominins have not benefited from any possible help, and their lifespan was significantly shorter than ours. Additionally, a hominin affected by a mental illness would risk turning from a skillful hunter/gatherer into easy prey.

Maybe, more importantly, an increased number of mentally sick individuals, who would turn in part disable, would gravely affect the social life of the small hominin communities known to run from 20-30 to 100 individuals.

To conclude, based on multiple and multidisciplinary pieces of evidence, I am confident that the main implications of the Laschamp event were intimately related to a typical high concentration of C 14 isotopes that always occurs during geomagnetic excursions.

In the meantime, there is very little evidence that the UV radiative effect was dominant and widespread.

Why prehistoric people migrated out-of-Africa?

I aim to show in this paper that Homo sapiens migration out of Africa can be correlated by some evidence suggested by neurogenesis and the Science of Complexity.

It is known that Homo sapiens existed in Africa in various regions. During the episode of severe cooling of the global climate, that occurred cca. seventy thousand years ago, the connections between these regions were interrupted. On top of global climate cooling, the catastrophic eruption of Mt. Toba (Indonesia) cca. seventy-four thousand years ago generated a type of regional volcanic cooling that lasted from 1,000 to almost 5,000 years.

We should consider the Norwegian-Greenland Sea Geomagnetic Excursion effects that ignited around 78,000 years ago, extending to 70,000 years ago. Here, the specialists have identified it differently in distinct regions of the Earth. Some estimates define it at a split event that occurred 70-67,000 years ago and again 57-54,000 years ago. But in Hexi Corridor (China) it was defined 80-70,000 years ago, and similarly in the Gulf of Bengal.

This excursion is thought to triggered global cooling.

In the meantime, this excursion is a generator of cosmogenic radioactivity in the form of a high concentration of atmospheric C 14 isotopes. The assimilation of C 14 by feeding causes a metabolic stimulation of oxidative processes and species (ROS). This ROS, before being deactivated by antioxidative processes, generated bursts of neurogenesis in Homo sapiens.

It occurred bursts of undifferentiated neurons, followed by their gradual differentiation. Such pulses of neurogenesis seem to be at the root of the first manifestation of mental abstractions, which are associated with an incipient art developed by the Homo inhabitants isolated in a climate refuge at the peak of South Africa (Blombos cave).

The neurogenetic effect of this geomagnetic event also seems to ignite some waves of migration out of Africa. The migrants benefited from mentioned neurogenesis bursts, generating undifferentiated stem cells, but these precursor cells' differentiation process was not fully developed. The undifferentiated bursts vastly increased the brain plasticity.

Hence, the migrants of these initial waves (70,000 years ago), and other earlier waves not discussed here (120,000-100,000 years ago), were not fully prepared to encounter new environments and various challenges. Their migration was intermittent in these early periods, while frequent local extinctions generated discontinue and interrupted settlements.

In this epoch (around 70,000 years ago), coastal Southern Asia's climate was not friendly. A dramatic change from 65,000 to 50,000 years ago occurred when the monsoon brought tropical and subtropical climate tendencies to such coastal areas.

Based on some archaeological findings, it could be determined that waves of Homo sapiens migration initially reached South India, South China, and Southeast Asia 70,000-65,000 years ago and New Guinea-Australia around 65,000-60,000 years ago.

Again, these migrants were not fully prepared to deal well with the areas they had to settle. I refer to two fundamental aspects: a weak cognitive development based on incomplete neural differentiation and a second weakness manifested against local pathogens and viruses.

Such mentioned situation of weakness caused frequent extinctions among various communities or groups, drastically limiting their environmental imprint.

This case reached archaeological notoriety when one compares the age of Indonesia's archaeological sites with those of New Guinea-Australia. The Indonesian ages of sites appear much younger than the New Guinean and Australian ones. In my opinion, it implies that the settlers in Indonesia perished at a fast rate, and their environmental imprint became almost undetectable.

On the contrary, the settlers of New Guinea and Australia showed a more robust behavior, and their settlements succeeded the trial of time. It may be the case that these people's brains encountered a better neural differentiation that helped with a more advanced adaptation.

It would be very speculative to explain why the settlers in the far east of this region (New Guinea-Australia) got reasonably accommodated to their environmental constraints. At the same time, those in the west (Indonesia) did not survive local conditions, while their early chains of extinction generated a 20,000 years negative imprint in their occupation of the territory.

However, the differentiation of neurons signified an increased capability toward various adaptations. The differentiation in the neural regions in charge of cognition (behavior, memory, learning, reasoning) provides a community of neurons that assembled complexities. The emergent results of such neuronal assemblies manifested as the intelligence of those neural systems.

I would say that the **burst of neurogenesis**, when the differentiation intervened, **were the creators of human intelligence**, while such neuronal processes did not manifest in any other species.

I documented these phenomena in several of my papers recently uploaded on academia.edu (like *Universal technology of language, Language brought us the intelligence of complexification*, and *The role played by the biological absorption of C 14 isotopes in stimulating neurogenesis and biophoton production*). Hence, I would not enter these aspects in-depth here.

Assumable, the burst of neurogenesis triggered changes in the brain anatomy and neural circuitry, shown by the skull rounding process. The skull rounding incipiently manifested in the last 70,000 years ago but turned more evolved only 35,000 years ago when a language revolution occurred, showing an operational language-ready brain.

The out of Africa migrants left their continent without a language-ready brain and very primitive speaking skills. The accomplishment of the mentioned issues was brought back to Africa only later by the Asian waves of reverse migration.

The new capability of neuronal differentiation became explicitly manifested around 65,000 years ago when migration from South Africa to Eastern Africa took place. East Africa's climate turned subtropical and tropical for the first time in 100,000 years, probably extending too to the coastal areas of Southwest Asia.

* * *

The migration 65,000 years ago signifies two things. The first thing is unveiled by the Science of Complexity that shows that the African evolution of Homo sapiens acquired a complexity level of "efficiency." Such a level prevented people from adapting to constant environmental changes. This conservative behavior forced them to choose small-style habitations in stable and fertile environments with relatively stable resources.

Hence, the African inhabitants find various climates and resources refuges on Africa's territory, like fertile valleys, or fertile lands around lakes, while these people became isolated within each sanctuary.

The neurogenesis bursts stimulated higher adaptability, especially among the inhabitants of the Southern African refugee, who were challenged by a more pronounced cooling around 70,000 years ago.

East Africa's tropicalization caused by a new path in the Indian Ocean monsoon opened a window of opportunity toward better feeding and more sustainable resources.

Hence, around 65,000 years ago, the Homo of South Africa started to extend into Eastern Africa, gradually traveling 500-800 km to Kenya, Somalia, Ethiopia. The same tropicalization occurred along the Arabian Peninsula's southern coastline, making fertile coastal lands to the Indian Peninsula. Crossing Bab-el-Mandeb Strait was like crossing a large African river.

* * *

It is essential to define <u>what migration meant and how worked the mechanism that produced it</u>.

Most specialists connect the size of hominin and Homo sapiens groups to certain features of the cortex. The mental ability to maintain relationships with other individuals limits the possible group size.

Based on archaeological findings, it is admitted that African groups were in the range of a minimum of 20 and a maximum of 150 individuals.

Around 70,000 years ago, the average African size of the group was 20-30 individuals. It is known that some size variation occurred between groups, while the group size increased over time.

The migration implies a particular mental state that directly and indirectly influences the <u>fertility rate</u> of the community.

When Homo sapiens were in South Africa, the fertility was low. But, around 65,000 years ago, that rate increased, which can be related to the mentioned bursts in neurogenesis. Better adaptability supplemented by the settlement in fertile territories increased the confidence of survival of larger communities and stimulated better reproduction rates. The increase was not in the group size but the rate of population doubling.

Let's see the mechanism that determines a species extension in the territory. As Professor Dunbar concluded in his extensive research on human evolution, the groups encounter a <u>doubling population split into two daughter groups</u>.

For a long time in Africa, the doubling always occurred, but it produced minimal territorial advances and tiny increases in inhabitants' number. In this case, the long length of doubling reflects a minimal rise in population. The group expansion was small, too, because, at the time, the environments previous to 70,000 years ago had plenty of resources.

For example, assumable a Paleolithic group needs a subsistence area with a radius of 125 km when the resources are on average. But, let's say along a fertile valley, the site of subsistence would be reduced ten times to a radius of 12 km.

The slow Homo expansion in Africa indicates that most groups inhabited very fertile areas for very long periods. The fertility here was low because the environmental constraints did not require a backup of higher fertility.

However, 65,000 years ago, the situation dramatically changed following climate change and the bursts of neurogenesis. It changed the way this mechanism regulated the behaviors of the group.

The fertility increase made the population's doubling occur in shorter intervals (around one generation). If doubling occurs every 20 years, the excess population over the stable 25 individuals must move away at a distance of 125 km from the center of the initial group's territory. Hence, in 20 years, the population increases in numbers, but also expends in the territory. In such a case, in 100 years, the people will linearly expand 625 km, in 1,000 years 6,250 km, and in 10,000 years some 62,500 km.

Considering a two-dimensional scenario: the initial settlement is A; the doubling makes people move east to B; now the doubling of A after another generation makes people move north to C; the doubling of B makes people continue to drive east to D; and so on. It shows that doubling makes people move in preferential directions, where the resources seem to fit the best. They populate the territory in a manner that is somehow uneven or discontinues.

However, the distance from Somalia to Australia, following the coastal landscape, is cca. 27,000-30,000 km and the groups from the above example would need 5,000-6,000 years to cover such a distance. In the meantime, the other daughter of the same population extends to the north, covering a similar distance. The entire globe can be populated in 12,000-15,000 years.

The pattern indicated above is a simple assumption because no information exists about Middle/Late Paleolithic fertility, mortality, and other critical data.

In the meantime, one should consider that some of the groups faced significant difficulties, which could range to extinction that produced spatial unevenness and disruption in our archaeological records.

Thus, the migration was not an actual migration process based on purpose. It occurred like a move for subsistence due to population increases.

The migrants did not know where they went or where they would go because they acted following a mechanism that implied <u>territorial expansion for subsistence</u> on whatever distance was required by the subsistence area's radius. But a radius may vary from one region to another, reflecting the availability or density of resources.

In every new areal of subsistence, <u>the group was forced to mentally adapt to the size of the problems they have to solve, advancing the neuronal differentiation that increased the cognition capacity</u>. **Experiencing mental complexification and producing emergent ideas led to the first cognitive intelligence** on this planet.

The migrants also genetically adapted to novel pathogens, viruses by forming new haplogroups.

All benefits produced by the out of Africa migration were brought back to the African population by back migrations.

Second Paper
Absorption of C14 isotope stimulates neurogenesis

Dan M. Mrejeru

The role played by the biological absorption of C14 isotope in stimulating neurogenesis and biophoton production

Abstract

This paper aims to demonstrate that low doses of ionizing radiation, occurring from C 14 isotopes, significantly influence human neurogenesis and produce biopositive effects.

This research may be relevant to social neuroscience, anthropology, psychiatry, nuclear medical technology applications, atomic safety regulations.

Low doses of ionizing irradiation elevate oxidative stress. The shift from a reduced state to an oxidized state acts as a cellular switch mechanism, affecting the stem cells and neurons' modality of action by moving them from proliferation to differentiation. In the first phase, the neurogenesis produces nondifferentiated new neurons, which are open to all possibilities and account for the process of plasticity. Plasticity serves to adapt to various environmental challenges. Later on, by complying with a particular adaptation, the neurons enter the differentiation phase that enlarges cognition. The proliferation slows down when the differentiation occurs.

A better-oxidized environment favors both cell proliferation and differentiation. The experiments found that proliferation or self-renewing affects the multipotent cell progenitors with a high ROS status. It means that cell proliferation and differentiation are highly responsive to ROS stimulation.

Superoxide dismutase (SOD) is a major antioxidant enzyme that removes superoxide radicals (free radicals).

Introduction.

Low-dose ionizing radiation is ubiquitous in our environment, and it corresponds to a dose of 100mSv or less. The dose is defined as a rate of radiation exposure of 6mSv or less per hour (Feng Ru Tang, 2016).

It should be noted that from 42,000 years ago to 2,500 years ago, humans have lived with 20-80% higher than standard atmospheric concentration C 14 isotope. Even when it occurred a type of intermittent exposure, the cumulative direct exposure to irradiation is estimated to a total length of over 10,000 years. Some other shorter events (like solar flares, solar minimum, and supernovae bursts) also contributed to ionizing radiation exposure.

Hence, it is of paramount importance to define the possible effect of such prolonged eras of low doses of ionizing radiation on human evolution. It is of particular concern for our brain development into a fundamentally novel intellect.

Let's have first a glimpse into current radiative conjecture being produced by our human-made activities.

C14 (carbon-14) is used in pharmacological and investigative research in carbonate form for isotopic labeling of molecules. Such activities use a dose greater than 1GBq, which means that the dose is around 0.4% (one in 250) higher than that provided by the natural amount of C14 isotopes in the atmospheric concentration. However, other pharmacological procedures indicate irradiation of 1 to 5 mCi compared to C 14 average irradiation at sea level known to be 2mCi per year in 1950 taken as reference for normality.

Among the medical procedures, CT scanning accounts for 40% of the population's annual irradiation exposure that uses such an approach for medical investigations.

However, C-14 is a low energy beta emitter, and even large amounts of this isotope pose little external dose hazard to persons exposed. The beta radiation barely penetrates the outer protective dead layer of the skin that covers the body. The primary concern refers to the possibility of internal exposure. The critical element for most C14-labelled compounds is the fat of the whole body. Beta radiation has a short-range in the air.

When C14 decays, it emits a beta particle and becomes an N-14 isotope (theoretically reactive) with a half-life of 7.5 seconds. All other isotopes, but not C14, have a half-life of fewer than 20 seconds.

C14 needs 5,730 years (half-life) to decay into a gamma particle and N14. After this decay, the remaining C14 and N14 are preserved within a 1:1 ratio.

A short view on dilution of C 14 isotope concentration in natural pools

There is a tremendous dilution of naturally produced C14 in the various pools of carbon in nature. The C14 is combined with CO_2 making a radioactive compound with an atmospheric half-life of 12-16 years in the northern hemisphere atmosphere.

C14 present in the atmosphere is mixed, and it is also mixed in shallow seawaters and freshwaters. In soils, as organically absorbed, it becomes associated with carbonate soil minerals.

According to the literature, the excess carbon 14 produced during nuclear weapons testing reached an 80% high in 1963-1965. From that peak, it has decreased due to the global carbon exchange cycle. By the 1990s, the carbon 14 level was only 20% higher than the theoretical 1950 level of reference. In 2020 that level will be again as in 1950.

As one can see, our exposure to C14 isotopes varied primarily over long and short periods. What was the possible effect of this variation?

Methods, materials, and results

I have searched Pub Med and other sources for English-language articles to acquire the necessary data needed to fulfill my discussion and interpretation.

The compilation allowed me to interpolate the introduced results. Such interpolation also provided the opportunity to compare data and bridge the stated effects. It led me to several conclusions, which have been distinct from other authors.

My research intended to evaluate the case of natural occurrences when the atmospheric concentration of C 14 isotopes exceeded typical values recorded before 1950. Such higher values (20-80% higher than usual) were comparable with today's nuclear medicine's clinical doses.

One of the most significant studies of Li-Chun Wei, Yiu-Xiu Ding, Yong-Hong Liu, Li Duan, Ya Bai, Mei Shi, and Liang-Wei Chen is *Low-dose radiation stimulates Wnt/beta-catenin signaling, neural stem proliferation, and neurogenesis in the mouse hippocampus in vitro and Vivo* published in the Journal of Radiation Research and in Current Alzheimer Research, January 2012.

Their study indicates that "Wnt/beta-catenin signaling is critical in the control of proliferation and differentiation rate of neural stem cells or progenitors in the hippocampus. In this study, the biological effects of low-dose radiation in stimulating Wnt/beta-catenin signaling, neural stem cell proliferation, and neurogenesis of hippocampus were interestingly identified by in vitro cell cultures and in vivo animal studies."

"First, low-dose radiation (0.3Gy) induced caused increasing of Wnt3a, Wnt5a, and beta-catenin expression in both neural stem cells and situ hippocampus by immunohistochemical and PCR detection."

"Secondly, low-dose radiation enhanced the neurogenesis of hippocampus indicated by increasing proliferation and neuronal differentiation of neural stem cells, going up of nestin-expressing cells and BrdU-incorporation in the hippocampus."

"Thirdly, it promoted cell survival and reduced apoptotic death of neuronal stem cells by flowcytometry analysis."

"Finally, Morris-water maze test showed behavioral improvement of animal learning in low-dose radiation group."

Another study of a group of Japanese researchers led by Norio Takahashi, Munechika Misumi, and Hideko Murakami is the Association between low doses of ionizing radiation, administered acutely or chronically, and *time to onset of stroke in a rat model*, published online 2020 Aug 4.

It came with an interesting definition of the radiation rate for biopositive and bionegative effects.

The authors found that rats acutely irradiated with doses between 0 and 1.9Gy or chronically irradiated with a cumulative dose of 0.5 or 1.0Gy (at a rate of 0.05 or 0.1Gy/day) indicated a threshold around 0.1Gy. Below the threshold of the mentioned low dose-rate but chronically exposed, no significant increase in stroke symptom was observed. The risk of stroke clearly appeared at high doses.

Feng Ru Tang and Weng Keong Loke's study is *Molecular Mechanism of low dose ionizing radiation-induced hormesis, adaptive responses, radioresistance, bystander effect, and genomic instability* in Int J Radiat Biol, 2015 Jan, pages 13-27.

They concluded that LDIR-induced hormesis is produced by P53 that directly seems to make the adaptive response, radioresistance, and genomic instability.

Feng Ru Tang and Konstantin Loganovsky wrote *Low dose rate ionizing radiation-induced health effect in the human*, published in

Elsevier Epub 2018 Jun 5. They concluded that LDIR or LDRIR exposure to irradiation might induce positive and negative effects.

Some other studies on animals found that a larger group of unregulated genes is not affected by low doses of 2Gy irradiation administered by radiotherapy. They found that some alterations in gene expression are qualitatively different depending on the dose being used.

The results indicate a significant opportunity to define the human brain's biopositive effect comparable with similar biopositive effects recorded during animal experiments.

When mice were exposed to low doses but for 20-30 generations, the results show a host of biopositive effects, like increased litter size, more fertile individuals than in the control group, increased litter number, increased viability, and faster growth rate.

The mice were exposed to 4.3mGy/day for three weeks. In another experiment, a mice colony was exposed for 21 generations to a 28.8mGy dose at 1.2mGy/h.

The lowest dose used was 31mGy, representing a radiation exposure of 4mCi, which produces an absorption of 0.4mGy/h.

Here, it should be noted that the dose of 4mCi is double compared to C 14 dose recorded in 1950 in the air over the sea level. In the meantime, this same dose equals the peak dose recorded for 10,000 years in our prehistory, and also it equals the peak dose recorded during the atmospheric atomic bomb experiments.

It has been observed that low doses of radiation, as used in the upper range of standard nuclear diagnostic, create even more efficient mutations than the much larger doses when reaching the DNA. Also, they produce an increase of 1.5-fold in the gene expression that could affect 6% of total gene loci.

The experiments recorded in the literature indicated that all bionegative aspects started to occur at doses much higher than 30mGy.

However, exploiting the opportunity to compare the results mentioned above from various studies and experiments could change our understanding and interpretation of exogenous irradiation's role (like C 14 isotopes) in ROS processes.

The materials used in this research refer to a collection of data obtained from other researchers' review work. As I showed in this section of the paper, an increase of 2mCi in the irradiation has produced significant biopositive effects in animal experiments.

It should be noted that human neurogenesis evolved differently than in the rest of animals because, in humans, the neuronal proliferation is directed from the Dentate Gyrus of the hippocampus to help neuronal plasticity (with undifferentiated neurons) and cognition (by differentiating the neurons). By contrast, in the rest of the animals, neuronal proliferation comes to support the olfactive function. One may say that much of animal reasoning is connected with olfaction.

The same process identically worked for humans until a fundamental turn took place. We do not know what prompted such a radical neuronal change or when and how this process took place.

Was it sudden or gradual?

Therefore, one can estimate much larger effects of the same low dose in human neurogenesis, which are preponderantly driven toward plasticity and cognition.

Discussion

The fluctuation of atmospheric C 14 concentration

Various research estimates that, even on a time scale of a few decades, the atmospheric radiocarbon activity may not have been constant.

"The atmospheric radiocarbon concentration fluctuated due to natural causes by 2% during the Little Ice Age from the 16th to 19th centuries". "The abundance of atmospheric radiocarbon (C14) from 1500 to 1800 AD was 2% above average level, with a maximum that occurred for the years AD 1500 and AD 1700, and a minimum for AD 1600 and AD 1800". "Between AD 1500 and 1850, two maxima, two minima, and other minor variations are known as de Vries effect have been observed. The C14 maxima correspond to the Maunder and Sporer minima of solar activity". "This effect modulates the galactic cosmic rays' activity in antiphase with solar activity." (I quoted from *Temporal fluctuation of atmospheric C14: causal factors and implications* by Paul E. Damon, Juan Carlos Lerman, and Austin Long with the University of Arizona, Tucson, published in Annual Reviews, Inc., 1978).

The global carbon cycle refers to the exchanges of carbon within and between four major reservoirs: the atmosphere, the oceans, land (lithosphere), and fossil fuels. Carbon may be transferred from one pool to another in seconds (by fixating atmospheric CO_2 into sugar through photosynthesis) or over millennia (the accumulation of fossil carbon through deposition and diagenesis of organic matter).

It is estimated (Nydal-1968) that the residence time in the stratosphere before transfer to the troposphere is 2.0 years. The transfer between Earth's hemispheres takes one year. The C14 atmospheric residence before transferring to the biosphere and aquatics is four years. Here, the residence in oceans is seven years, and from the ocean surface to the deep, the transfer takes 24 years.

Lingenfelter and Ramaty (1970) calculated that the solar flare of February 23, 1956, may have increased the atmosphere's radiocarbon concentration by 7.5%.

Paul A. LaViolette estimated in 2011, based on observations in the Cariaco Basin out of Venezuela coast, that a giant cycle of solar flares occurred in phases around 12,973 years ago, 12,837, and 12,639 years ago. It extended on a length of about 300 years. The radiocarbon concentration rose by more than 50%.

A much smaller occurrence was the Carrington event of 1859, but which was still 20 times greater than the 1956 event.

Another estimate indicates that Global Warming would cause a decrease in atmospheric C14 concentration.

"Because the industrial CO_2 injected into the atmosphere has no C14, the effect decreased the C14 atmospheric concentration. This effect corresponds to the Industrial Era. It is estimated that C14 concentration before the industrial era was 2% higher than in 1950."

Konstantinov and Kocharov (1967) estimated a 0.5 to 40% increase of radiocarbon due to the gamma-ray burst from the supernova Tycho Brahe, seen in AD 1572.

Most such measurements were made on tree rings. The oldest bristlecone pine chronologies extend to 8,400 years ago.

The exchange time between the mentioned three reservoirs is ten years.

While the absolute sum of carbon found in the active reservoirs is maintained in a near steady-state by slow geological processes, more rapid biochemical processes drive carbon redistribution among the active reservoirs.

Dr. Zalasiewicz explains: "As a result of the detonations of hundreds of such weapons (atomic) around the globe, there will be plenty of these isotopes still around far into the future." (He refers to inorganic matter).

The amount of fossil fuel combustion decreases the C14 because fossil fuels have lost, on long geological scales, all C14 because of radioactive decay. In contrast, such fuel emission adds large nonradioactive volumes (CO_2).

The biological implication of changes in the gaseous concentration

It is considered that today C14 concentration is 2% enriched compared to the preindustrial era (from over 200 years ago). The current trend in emissions continues to decrease the C14 concentration that, around 2020, it will reach the level from the preindustrial era.

By 2040, the concentration of atmospheric C14 will diminish by one quart compared to the preindustrial era.

Here, I like to note that such a decrease, from double concentration (around 1963) to 25% less than the concentration assumed to exists in the preindustrial era, may have vast mental implications.

However, such a change in C14 atmospheric concentration is generated by increasing to double the carbon dioxide during the same time interval.

It is well-known that it exists a direct proportionality between these gases concentration in the atmosphere: one component, like the carbon dioxide, rises, the other elements, like nitrogen (N), Oxygen, and C14, would diminish correspondingly.

For example, studies of euthanasia on mice have revealed the influence of such euthanasic gases on mice subjects:

-CO_2 use increased the locomotor activity;

-N_2 decreased such locomotion;

-CO2 increased the delta, theta rhythm correlation on the EEG, and increased excitation;

-N2 reduced behavioral activity and central neurological depression while being less aversive.

Both gases act on the brain and nervous system.

Other studies were concentrated on the effects of pollution on our neurological system. I will note several found effects:

-traffic-related air pollution (with mainly the carbon dioxide) exposure was associated with adverse effects, like a reduction in attention, global IQ, memory, and higher prevalence of several mental diseases (ADHD, ASD);

-it was observed a cognitive decline and increased dementia behaviors.

Global atmospheric CO2 levels passed 400 parts per million in 2016, double compared to the preindustrial era.

As the studies point out, the CO2 concentration in the atmosphere tends to increase almost to 1,000 parts per million toward the end of the current century.

However, we have frequent cases where the indoor concentration of CO2 is daily higher than 1,000 ppm and even higher than 2,000 parts per million, especially in the school environment and almost all office spaces. Even the concentration can reach over 1,000 parts per million in our houses when they are tightly isolated from the outdoor environments for energetic preservation reasons.

In sum, there are significant changes in the gaseous concentration in our planetary atmosphere, where such changes affect our biology. Most importantly, they affect our brain workings in a very substantial manner.

It is hard to imagine that our Late Paleolithic brain would accommodate such functional changes when the previous plasticity source, like C14 high atmospheric concentration, diminishes significantly.

One last aspect is global nitrogen pollution on the rise, but this subject does not enter the issues I like to discuss here.

The role of C 14 in the mechanism of our biology

Let me continue with the role being played by C14 in the evolution of the human brain.

The animal organisms assimilate C14 Glucose when feeding on vegetables. This process stimulates the oxygen species (ROS) and regulates nitric oxide production and inhibition (NO), which have interrelated neurogenesis roles.

Under certain circumstances, experimental animal data suggest that the exposure gives anti-tumor ability, slows the progression of arteriosclerosis, and ameliorates diabetic nephropathy.

However, during experiments, the C14 traces have been 1 to 5mCi, thus entering the low hazard category.

The average concentration of C14 isotopes at sea level is 2mCi per year. During the atomic bomb testing in the atmosphere, such C14 concentration rose to 3.6mCi (80% higher than usual) as it was investigated in 1963-1965, and it diminished to 2.4mCi in 1990 (20% higher than average).

Thus, the atmospheric data mentioned above could be correlated with the dose used in pharmaceutical and medical experiments and procedures. The length of the exposure would produce fundamental differences.

The C14 isotopes inside the human body, as part of C14 Glucose, would stimulate the production of oxygen species (ROS), which have specific reactions with nitric oxide, regulating its concentration in various tissues. NO is known to create a nitrite reduction pathway. Within the blood vessels, the Oxygen produces vasoconstriction that causes hyperbaric tension within the vessels. High tension is also found inside the cells, inflicting the level of permeability of the cellular membranes. It increases the amount of available Oxygen in blood vessels and cells and interacts with local nitric oxide, regulating its production.

Nitric oxide is involved in transporting and dispersing Glucose in the entire organism. Glucose also stimulates oxygen species that metabolize any unwanted excess of this substance. When nitric oxide (as an antioxidant) decreases in the brain and nervous system, local oxygen species increase. By contrast, the same vascular tension affects the vascular peripheries distinctly, making the nitric oxide increase locally (vasodilation) while stimulating muscle fibers and skeletal developments.

The ROS regulates cellular differentiation, proliferation, apoptosis, cell cycle, and migration.

It is known that, during physical activity, the ROS and NO interplay stimulates oxygen production.

The effect of low doses of radiation

The total amount of ROS generated by primary ionizing events (like C14 isotope activity when inserted into C14Glucose) is propagated to the level of intracellular activation. Almost immediately after exposure to ionizing radiation, the detoxifying enzymes activate the antioxidant system's cellular defense mechanism. An entire biological mechanism converges to repair nucleic acid damage and DNA modifications induced by the ionizing radiation, like the one generated by the atmospheric C14 isotopes inserted into the Glucose produced by plants.

Low-dose Ionizing Radiation (IR) at 15-30mGy induces in biological molecules Reactive Oxygen Species (ROS) because gamma radiation of cellular water rapidly generates ROS like hydroxyl radical (OH) and ionized water.

The literature analyzed experimental cases where the irradiation of A549 cells induced mitochondrial ROS production, increased mitochondrial membrane potential and promoted respiration and ATP production.

The experiments with doses lower than 30mGy represent 4mCi radiation exposure, which produces an absorption of 0.4mGy/hr. They proved to have no statistical significance. On the other hand, it may increase gene expression and trigger higher than regular mutation rates.

Experiments on mice colonies were prolonged to several generations, and the dose used was 4.5mGy/day to 28.8mGy at 1.2mGy/h. These experiments recorded many biopositive effects, like increased litter size, more fertile than the control group, increased litter number, increased viability, and faster growth rates.

Production of biophotons

However, a chemical reaction in cells caused by oxidation produces free radicals, which are the primary source of biophotons. Such biophotons are made during the free radicals deexcitation process.

Brain plasticity refers to an undifferentiated neuronal state. When the entropy occasionally increases in the hippocampus's Dentate Gyrus, a neuronal generation will burst with higher than average plasticity, meaning more undifferentiated neurons. It is known that higher oxidative stress generates a higher entropy.

Scientific literature estimates that biophotons would play a significant role in neuronal transmission and communication. Hence, the ionizing radiation inserted in C14 Glucose generated oxidative stress that resulted in a high biophoton production. These steps contributed to anatomical and functional changes in the Homo brain, producing a new type of mental activity and a modern intellect. The ultimate result was a revolution in language prompted by a language-ready brain.

Human neurogenesis

It has been experimentally demonstrated that a high concentration of atmospheric C14 has contributed to increased neurogenesis and higher brain plasticity by combining the interpolations.

The current diminishing atmospheric concentration of C14 seems to have implications for future human generations, making them less adaptable to changes.

This is important because we enter an Artificial Intelligence implementation era, where we will need a significantly augmented capacity of adaptation.

Another issue to pay attention to is the ingestion of inorganic minerals by organisms, including humans.

Dr. Rick Wagner, C.N., explains in his paper Ionic Minerals:

"In nature, our mineral sources have always been the water-based either form of foods we eat or the water we drink. When minerals come from water, they are inorganic. When one ingests sea salt for the mineral content, they are utilizing inorganic forms of the minerals ".

"The important thing is that it does not matter if you consume minerals from plant sources, water-based, or solid pills or even dirt. What matters is how much time it takes the body to break down the minerals into their atomic (ionic) state to be effectively utilized at the cellular level. The body does not need its minerals bonded to carbon (ions are not bounded to carbon) for beneficial absorption and utilization. A mineral is an ionic form that alleviates the need for stomach acid to perform the function of ionization".

There is known that C14 usually bounds with sodium, potassium, and magnesium.

Investigation of C 14 trackers

Returning to the investigative-use of C14 tracers, there is known that the human organism has a safety system that, in less than 48 seconds, eliminates a source that posits biological danger.

Such a trace-use of C14 isotope, as a beta emitter, is low-dose equivalent and has insufficient energy to cause ionization but has enough energy to produce local excitation.

A radiotracer is a radioactive element, like the C14 isotope, added to a nonradioactive part.

Various pharmacological testing and experiments with isotopic C14-labeled products (mainly C14 labeled glucose) resulted in an electrical stimulation that doubles the rate of glucose consumption. They showed a trifold increase in the rate of CO_2 production.

These experiments indicated that the required energy increased as a direct result of increased respiration and glycolysis. Here, a supplementary metabolism was present, generating Oxygen, which was necessary to satisfy the increase in oxygen consumption.

However, the changes in the absorption of C14-labeled products by various cerebral structures resulted from their activation and the isotopic incorporation rate. When one substrate was involved in the mechanochemical reaction, the resultant product had the same isotopic composition as the substrate. It was found that C14 is stably into organic molecules and blood. Of interest to me, there was an increase of C14-glucose in the hypothalamus that generated an acute activation at the beginning of the experiments that was followed by a diminishing activation as a specific adaptation occurred.

Beyond the medical interest in the use of C14 as investigative tracers, it must be considered some other aspects of C14 isotope absorption by biological systems.

Natural and artificial cause of increases in C 14 concentration

I have to specify that cosmic radiation, which produces atmospheric C 14 isotopes, currently generates a level of 0.3mSv per year at sea level.

During the atomic bomb atmospheric experiments, the mentioned above level was almost doubled to 0.5mSv. The same doubling occurred during the geophysical events produced by the geomagnetic excursions.

A person flying at 10,000 meters above the ground for 4 hours is subject to 20microSv. By comparison, a tracer used in medical and pharmaceutical fields can cause exposure of 0.2 to 5.2microSv, which is four times lower than the mentioned flight exposure.

When C14 decays, it emits a beta particle. The beta particle generates free-radicals. The presence of free radicals or reactive oxygen species (ROS) corresponds to ultraweak photon emission (UPE). ROS production plays a crucial role in defense against infections, apoptosis, aging, and cellular communication. The presence of reactive nitrogen species (RNS) that is the second element (as N14) encountered during C14 decaying also busts UPE or biophoton production.

However, one should consider that since a high atmospheric concentration existed for almost 10,000 years, such concentration (20% to 80% higher than usual) was preserved in inorganic systems at the same level for 6,000 years, each time it was produced. The last known high concentration of atmospheric C14 was produced by the Sterno-Etrussia geomagnetic excursion of the geomagnetic field 2,500 years ago. It would remain present at its initial concentration in inorganic forms for 3,500 years from now on. It may continue to be absorbed from inorganic into biological systems.

There is information that neural theta rhythm (central to language production) is correlated with biophoton emissions.

The research has found that the biophotons are sent along with the axonal and neuronal ramifications and throughout the nervous system. Such biophoton conduction occurs with greater emphasis on the white matter of the brain representing the communication channels between the gray matter structures that define the neurons. Reactive oxygen and nitrogen species (ROS and RNS) are also part of the cellular metabolism and have beneficial effects. The balance between prooxidants and antioxidants is critical for the survival and function of (aerobic) organisms.

Oxygen and neurogenesis

The following part of this paper is dedicated to what the literature calls the critical role of Oxygen in regulating cell behavior during neurogenesis.

It intends to bridge the natural occurrence of high atmospheric concentration of C14 in our recent prehistory, producing a significant stimulation inflicting human neurogenesis.

Joseph Tafur, M.D., Eduard P.A. Van Wijk, Ph.D., Roeland Van Wijk, Ph. D., and Paul J. Mills, Ph.D. in their paper "*Biophoton Detection and Low-Intensity Light Therapy: A Potential Clinical Partnership*," published in Photomedicine and Laser Surgery, Volume 28, Number 1, 2010, Pp. 23-30, made a review of current knowledge on this matter. Some of their conclusions serve very well the purpose of my investigation.

"We know that Low-Intensity Light Therapy (LILT) has the potential to accelerate ATP production and mitigate oxidative stress, which derives from excessive production of reactive oxygen species (ROS) or a lack of antioxidant activity."

"Cellular sensitivity to red and infrared light is influenced by the cellular redox state. The cellular growth phase, which may also correspond to the cellular redox state, appears to be another determinant of this sensitivity. Proliferating cells are, in many cases, more sensitive. In each case, this proliferating phase is associated with elevated ROS production".

Here, the authors indicate that the proliferation of cells, neurons included, depends on the oxygen species' effect or redox. These effects are oxygen-dependent and involve the generation of ROS.

Wulf Droge, in his paper "*Free Radicals in the Physiological Control of Cell Function,*" published online on January 1, 2002, indicates:

"At high concentration, free radicals and radical-derived, non-radical reactive species are hazardous for living organisms and damage all major cellular constituents. At moderate concentrations, however, nitric oxide (NO), superoxide anion, and related reactive oxygen species (ROS) play an important role as regulatory mediators in the signaling process. Many of the ROS-mediated responses protect the cells against oxidative stress and reestablish redox homeostasis".

Here, it is essential to note that the C14 isotope, when its absorption occurs in a biological system, tends to decay into nitric oxide (NO) that would be lethal, as most scientists have previously thought. However, this old thought significantly changed in recent years. "Higher organisms, however, have evolved the use of NO and ROS also as signaling molecules to other physiological functions. NO, and ROS are typically generated in these cases by tightly regulated enzymes such as NO synthase (NOS) and NAD(P)H oxidase isoforms".

"In a given signaling protein, oxidative attack induces either a loss of function, a gain of function, or a switch to a different function. In mitochondria, ROS are generated as undesirable side products of oxidative energy metabolism. The process of aging may result, at least in part, from radical-mediated oxidative damage".

"Free-radicals are important in biology because of their advantageous effects. The role of nitric oxide (NO) was discovered as a regulatory molecule in the control of smooth muscle relaxation and platelet adhesion inhibition. A large body of evidence shows that living organisms have not only adapted to an unfriendly coexistence with free-radicals but have developed mechanisms for the advantageous use of free-radicals. The delicate balance between the advantageous and detrimental effects of free-radicals is an important aspect of life".

I quoted this author because he indicated how biology adapted to the production and presence of free-radicals.

Further, one will see that biology developed precise systems and reactions capable of de-excite or annihilate the free-radicals immediately after their mission was accomplished advantageously. Such a process prevents the occurrence of damages.

In the meantime, both papers quoted above indicate the stimulant. At the same time, the vital role of the oxidative stress that goes from increasing the cell/neuron proliferation (like in neurogenesis) to increased activation and signaling to the production of free radicals that, among other things, it seems to be the sole mechanism in charge with biophoton emission.

In an article ("*Oxygen, a Key Factor Regulating Cell Behavior during Neurogenesis and Cerebral Diseases*") written by Kuan Zhang, Lingling Zhu, and Ming Fan, and published in Frontiers in Molecular Neuroscience, April 4, 2011, the authors explain:

"Oxygen is a significant substrate for energy production and cell metabolism. It is interesting to note that normal oxygen levels in the tissues are always substantially lower than 156 mmHgO2 in the air we breathe (Panchison, 2009)."

"The brain is one of the heaviest oxygen consumers in the body, which amounts to 20% of total oxygen consumption (Masamoto and Tanishita, 2009). The development of various organs of embryos, including the central nervous system (CNS), occurs in low-oxygen concentration

(Fisher and Bavister, 1993. Chen et al., 1999). Apart from this, oxygen levels in brain tissues are often altered during stroke (Liu et al., 2004), brain trauma (Valadka et al., 1998), and in hyperbaric oxygen (HBO) environment (Balenane, 1982)".

However, hyperbaric oxygen therapy (HBOT) refers to Oxygen's medical use at a level higher than atmospheric pressure (Jun Mu and John H. Zhang).

"Embryonic neurogenesis begins at the early gestation period under very low-oxygen concentration (15.2mmHg; Zhou, 2004)". "Taken together, all of above researches indicated the significant role of the vessels as one of the important components of the neurogenesis niche, and may also imply that the higher oxygen tension around the vessels in the subventricular zone (SVZ) and Dentate Gyrus (DG) would be significant for the maintenance of the characteristics of NSCs."

"We found that the PO2 (O2-oxygen) levels in ventricles are in a dynamic state and fluctuate in the range of 42 to 48 mmHg at a frequency of about 3 minutes. In the hippocampus, the PO2 level in CA1 and hilus are very stable and maintain about 2mmHg, while the PO2 (Oxygen) in Dentate Gyrus is dynamic and fluctuates in a range of 6-8mmHg (Zhang et al., 2010)".

"The PO2 levels in uninjured brain tissue have been measured about 25-30mmHg in the white matter of frontal lobe (Sarrafzadeh et al., 1998), 20-40mmHg in normal tissue (Hlatky et al., 2003)".

"Proliferation was promoted, and apoptosis was reduced when cells were grown in lower O2, yielding a greater number of precursors. The differentiation of precursor cells into neurons with specific neurotransmitter phenotypes was also significantly altered".

"In summary, the moderate low-oxygen (15.2-38mmHg) concentration was able to promote the proliferation of NSCs from various resources and enhance the differentiation of NSCs into the TH-positive neurons. It was found that the metabolism of cells would consume the Oxygen of the medium".

I like to quote Amandeep Goyal, Tyler Chonis, and Jeffrey S. Cooper with their paper "*Hyperbaric Cardiovascular Effects,*" published by Stat Pearls Publishing, January 2019.

"HBOT initiates generalized vasoconstriction of healthy blood vessels. Exposure to Oxygen pressure of at least 2 ATA is known to induce arteriolar vasoconstriction and increase systemic vascular resistance. The primary mechanism that leads to this vasoconstriction involves a reduction of nitric oxide production in the endothelium. The hyperoxic environment leads to increased oxidation of nitric oxide (NO) radicals produced by the endothelium, leading to a loss of the vasorelaxant effect. Additionally, some research has shown that HBOT leads to alterations in other vasodilator compounds, such as prostaglandins, contributing to the net vasoconstriction effect.

Hyperoxia stimulates the sympathetic nervous system to promote vasoconstriction. Humans have been shown to have an augmentation of sympathetic nervous system activity."

"Although vasoconstriction partially impedes blood flow, the hyperoxygenation of the plasma results in an overall gain in delivered Oxygen. Short-term hyperoxia causes increased cerebral vasoconstriction and further the reduction of blood flow. However, even with the reduction of cerebral blood flow, the cerebellum receives more Oxygen than it would otherwise".

"The brainstem neurons play important roles in the cardioinhibitory center and underlie the hyperbaric reflex bradycardia. HBOT leads to parasympathetic activity and increases vagal tone. It has been suggested that another factor leading to the slowed heart rate is a nitrogen-dependent beta-blockade of the heart."

"Reduction of blood flow secondary to vasoconstriction leads to corresponding edema reduction."

C 14 variability during geomagnetic excursions and the biological effect

Now, I like to connect the information about C14-labeled products and naturally occurred atmospheric showers of C14 isotopes due to the effect caused by the diminishing of the geomagnetic field intensity.

The geological samples, coral, tree-rings, and several other sources provide a record of the concentration of C14. In the past 42,000 years, our planet was subject to a high-frequency of short-bursts of gamma radiation, followed by showers of radionuclides, which we're incorporating the mentioned C14 isotope.

The C14 on the ground was absorbed by plants' photosynthesis and passed to animals and humans. Such absorption/incorporation process is similar to making C14-labeled products used as tracers for medical investigations. In the meantime, the C14-labeled products proved to act as biological stimulants, increasing metabolism and oxygen consumption. Such "increased oxygen production" from C14-labeled products was similar in many respects to "hyperoxia" also produced in the medical field.

In short, the release of Oxygen by neuronal processes exposed to the absorption of C14 isotopes generated a hyperbaric oxygen condition (vascular constriction).

Paul M. Macey, Mary A. Woo, and Donald M. Harper, of University College of London, United Kingdom, published an article titled *"Hyperoxic Brain Effects Are Normalized by Addition of CO2"* PLOS/Medicine on May 22, 2007.

The authors show that "despite the objective of improving the tissue oxygen delivery, hyperoxic ventilation could accentuate ischemic trouble and impair the outcome. But several cortical, limbic, and cerebellar brain areas regulate these autonomic processes".

"We found, using functional magnetic resonance imaging, that 2 minutes of hyperoxic ventilation (100% O2) following a room air baseline elicited pronounced responses in autonomic and hormonal areas, including the hypothalamus, insula, and hippocampus, throughout the challenge".

"The addition of 5% CO2 to 95% O2 abolished the responses in the hypothalamus and lingual gyrus, substantially reduced insular, hippocampal, thalamic, and cerebellar patterns in the first 48 seconds, and abolished signals in those sites after that. Only the dorsal midbrain responded to hypercapnia, but this would be abolished, too, by additional CO2".

At the beginning of this paper, I indicated that the pharmacological research with C14-labeled Glucose experimentally produced a twofold reduction of Glucose and a trifold increase of CO2.

Here, a neural mechanism was unveiled, which automatically, within 48 seconds (to two minutes), would naturally reduce the effects of hyperoxia (generated by C14-labeled products and/or hyperbaric therapy).

It should be assumed that, during our recent prehistory, the same neural mechanism was in charge of reducing the natural hyperoxia produced at the time in the brains of our ancestors by the absorbed C14 present in their food and derived from a high concentration of this isotope in the terrestrial atmosphere.

Such a safety procedure was built-in into our neural network. It reduced the mentioned encounter to a series of short-term pulses, and where such short-pulses prevent the development of free-radicals.

I indicated that the concentration of atmospheric C14 isotopes was varying from 20-30% to 70-80% higher than usual for a period of almost 30,000 years in our recent prehistory (from 42,000 years ago to 2,500 years ago).

Conclusion

However, now I would assume that the past-occurrence of a high concentration of C14 isotope was producing a cerebral pulsing-process with durations of less than 48 seconds.

Such pulse-process, developed over 10,000 years, was the natural stimulus affecting increased neurogenesis that shaped our modern brain into producing the plasticity necessary to create that language with which was buildup our civilization.

Here, it should be noted that an increase in the oxidative-stress produces free-radicals; the deexcitation of free-radicals generates biophotons. In the case of C14 isotope infliction, the neuronal absorption process also produces oxidative-stress. And this last process, where the oxidation is termed hyperoxic, produces vessel constriction that mounts hyperbaric pressure; here, the nitric oxide radicals (NO) appear when the oxidation is longer than 48 seconds. The problem is resolved by a vessel relaxation that follows due to a subsequent increase in CO_2 production in the brain.

There is possible that the C14 isotopes could generate oxidative stress, while two distinct processes annihilate the stress:

-the deexcitation of radicals that produces biophotons;

-increased local production of CO_2 that relaxes the constricted vessels, terminating the hyperbaric stress and its oxygen production. Suppose the hypothesis of a single cause for oxidative-stress could be proved. In that case, it may result that the same C14 infliction, in a particular era in our prehistory, generated simultaneously but developed in two distinct processes, increased neurogenesis, and separately, it increased the biophoton production.

Here, the biophoton production can represent the source of a proposed neural optic communication. It could also serve as a fundamental for imaginary thinking that is activated during entanglement.

I like to bring to this end several quotes from two recent papers.

One of these papers indicates strong evidence that "*Long Course Hyperbaric Oxygen Stimulates Neurogenesis and Attenuates Inflammation after Ischemic Stroke*" (a paper of a large team of researchers from National Cheung Kung University, Kaohsiung University, Southern Taiwan University, all of them in Taiwan, published by Hindawi Publishing Corporation, Volume 2013).

The second paper shows that "*Hyperbaric oxygen therapy promotes neurogenesis*" (written by a research team from Loma Linda University, US, and Chongqing Medical University, China, published online 2011, June 27).

In the first paper, the authors said: "Several studies provided experimental evidence concerning the neuroprotection benefits of hyperbaric oxygen (HBOT) therapy and the influence of this therapy on the migration of BMSCs, neurogenesis, gliosis, and inflammation."

The second paper suggests that hyperbaric oxygen therapy enhances neurogenesis. The authors demonstrate the influence of HBOT on cellular transcription factors, including hypoxia-inducible factors.

To conclude, I would suggest a potentially influential pathway of a high concentration of atmospheric C14 with a significant impact on our brains in prehistoric times.

Atmospheric C14 is assimilated by plants during photosynthesis, helping the sucrose formation that is half Glucose and half fructose. Further on, in this case, the Glucose becomes animals ingest the C14-Glucose along with the plants they consume.

In the human case, high C14-Glucose must increase nitric oxide production because it reduces blood glucose by uptake to skeletal and muscle systems. Ultimately, after a pulse of increased production, it occurs a general reduction of nitric oxide in some of its components.

Lesser nitric oxide generates vasoconstriction that increases the cells' oxygen tension and increases neuroblast proliferation (neurogenesis). Vasoconstriction in the brain stimulates neurogenesis, and consequently, brain plasticity and cognition.

The decrease in some nitric oxide components produced an increased resistance to viruses and pathogens. Such novel immunity helped with the migration out of Africa and adaption to new ecological conditions.

Acknowledgment

This paper's scope was to collect information on the exogeneous conjecture of geophysical forcing events that drove the human brain into a fundamental while unique transformation, giving rise to a language-ready intellect that made us distinct from anybody else on this planet.

This paper provides the core support for my hypothesis on the emergence of Homo loquens species just right out of the Homo sapiens precursors.

References

E. A. Ostrakhovitch, O. A. Semenikhin, *The role of redox environment in neurogenic development,* Elsevier June 2013

Ting-Tang Huang, Yani Zon, Rikki Corniola, *Oxidative stress and adult neurogenesis-effects of radiation and superoxide dismutase deficiency*,

Semin Cell Dev Biol, 2012 Sept 23: 738-744, published Online April 2012

Janet E. Le Belle, Nicolas M. Orozco, Andres A. Paucar, Jonathan P. Saxe, Jack Mottahedeh, April D. Pyle, Hong Wu, and Harley I. Kornblum, *Proliferative Neural Stem Cells Have High Endogenous ROS Levels that Regulate Self-Renewal and Neurogenesis in a*

P13K/Akt Dependent Manner, published in Cell Stem Cell, 2011 January 7, pages: 59-71.

Julie A. Reisz, Bidhi Bansal, Jiang Qian, Weiling Zhao, and Cristina M. Furdui, *Effects of Ionizing Radiation on Biological Molecules- Mechanism of Damage and Emerging Methods of Detection*, published in Antioxid Redox Signal, 2014 Jul 10, pages 260-292

Mohamed Ariff Iqbal and Eftekhar Eftekharpour, *Regulatory Role of Redox Balance in Determination of Neural Precursor Cell Fate*, published in Stem Cells International, Special Issue: Environmental Stimulus on Stem Cells Behavior, published 18 Jul 2017

Stanton Segal, Mones Berman and Alberta Blair, *Metabolism of variously C14 labeled Glucose in man and estimation of the extent of Glucose metabolism by the hexose monophosphate pathway*, published March 9, 1961, by National Institute of Arthritis and Metabolic Diseases, Bethesda, Md.

Paul E. Damon, Juan Carlos Lerman, and Austin Long, *Temporal Fluctuations of Atmospheric C14: Causal factors and Implications*, in Ann. Rev. Earth Planet Sci 1978 pages 457-94

Paul A. LaViolette of The Starburst Foundation, *Evidence for a Solar Flare of the Pleistocene Mass Extinction*, published in Radiocarbon January 2011

A group of researchers led by S. Klose, S. Schmidt, D.A. Kann, A. Nicuesa Guelbenzu, and A. Schulze, *Four GRB-Supernovae at Redshifts between 0.4 and 0.8*, published in Astronomy and Astrophysics manuscript no. ms August 2018

William G. Kaelin Jr., Sir Peter J. Retcliffe, and Greg L. Semenza, Press release: The Nobel Prize in Physiology or Medicine (2019 10-07 The Nobel Assembly Karolinska Institutet), *How cells sense and adapt to oxygen availability.* Daniel E. Duggan and Elwood O. Titus, with a study titled as *The use of radioactive isotopes for pharmacological research*, published by Springer-Verlag Berlin-Gottingen-Heidelberg 1961

Feng Ru Tang, Weng Keong Loke, and Boo Cheong Khoo with the study *Low-doses-rate ionizing radiation-induced bioeffects in animal models* published by Journal of Radiation Research and J Radiat Res, 2017 Mar, pages 165-182

Susan C. Tilton, Lye Meng Markillie, Spencer Hays, Ronald C. Taylor, and David L. Stenoien published the study *Identification of Differential Gene Expression Patterns after Acute Exposure to High and Low Doses of Low-LET Ionizing Radiation in a Reconstituted Human Skin Tissue* in Radiat Res 2016 Nov pages 531-538

Feng Ru Tang and Konstantin Loganovsky had the study *Low Dose rate ionizing radiation-induced health effect in the human* published by Elsevier, Epub 2018 Jun 5

Norio Takahasi, Munechika Misumi, Hideko Murakami, Yasuharu Niwa, Waka Ohishi, Toshiya Inaba, Akiko Nagamachi, and Gen Suzuki published the study *Association between low doses of ionizing radiation, administrated acutely or chronically, and time onset of stroke in a rat model*, published online 2020 Aug 4 by Journal of Radiation Research

Li-Chun Wei, Yin-Xiu Ding, Yong-Hong Liu, Li Duan, Ya Bai, Mei Shi, and Liang-Wei Chen had the study *Low-dose radiation stimulates Wnt/beta-catenin signaling, neural stem cell proliferation and neurogenesis of the mouse hippocampus in vitro and in vivo* published by Research Gate in January 2012 at Current Alzheimer research, pages 278-89

A research team led by Mari Katsura, Hiromasa Cyou-Nakamine, Qin Zen, and Yang Zen had the study Effects of Chronic Low-Dose radiation on Human Neural Progenitor Cells, published by Scientific Records (article number 20027-2016) on January 22, 2015

Tzu-Chien, Richard G. Fairbanks, Li Cao, and Richard A. Mortlock had the study Analysis of the atmospheric C 14 record spanning the past 50,000 years derived from high-precision Th 230, U 234, U 238, Pa

Seven Papers

231, U 235, and C 14 dates on fossil corals, published by Elsevier (Quarterly Science Review 26) on June 19, 2006

Kuan Zhang, Lingling Zhu, and Ming Fan had the study *Oxygen, a Key Factor Regulating Cell Behavior during Neurogenesis and Cerebral Diseases*, published in Frontiers in Molecular Neuroscience, April 4, 2011

Wulf Droge with the study *Free radicals in Physiological Control of Cell Function* published online January 1, 2002

Joseph Tafur, Eduard P. A. Van Wijk, Roeland Van Wijk, and Paul J. Mills with the study *Biophoton Detection and Low Intensity Light Therapy: A Potential Clinical Partnership*, published in Photomedicine and Laser Survey, volume 28, number 1, 2010, pages 23-30

Amandeep Goyal, Tyler Chonis, and Jeffrey S. Cooper with the study *Hyperbaric Cardiovascular Effects*, published by Stat Pearl Publishing, January 2019

Paul M. Macey, Mary A. Woo, and Donald M. Harper with the study *Hyperbaric Brain Effects are Normalized by Addition of CO_2*, published in PLOS/Medicine, May 22, 2007

Chun-Xia Luo, Xing Jin, Chang-Chun Cao, Ming-Wei Zhu, Bin Wang, Lei Chang, Qi-Gang Zhou, Hai-yin Wu, and Dong-Ya Zhu with study *Bidirectional regulation of neurogenesis by neural nitric oxide synthase derived from neurons and neural stem cells*, published by Stem Cell 2010 Nov, 28 (11) Prehistoric 2041-52 10.1002/stem.522

Third Paper
A language – ready brain

Changes in brain anatomy inflicted the development of language.

It is postulated, by several researchers, that the emergence of our species type of language-ready-brain is caused by developmental changes of brain morphology and neural connectivity. These changes give our braincase a more spherical shape.

I. Tattersall (April 2, 2019) suggests within his "*Minimalist Program and the Origin of Language*" (published in Frontiers) that language emerged in a short-term event that occurred abruptly and late. He suggests but also finds it debatable that "language was externalized in an independent event that followed its initial appearance."

My paper's focus refers to the fact that no archaeologist found an exact globular braincase but dated before 35,000 years ago. Thus, I try to demonstrate here that the anatomical changes and alterations in the brain connectivity resulted from a particular stimulus developing 42,000 years ago by a high atmospheric concentration of the C14 isotope. This stimulus has been present in the Earth's atmosphere from 42,000 to 2,500 years ago, which covers an effective influence developed intermittently over a total interval of 10,000 years.

However, geophysicists are aware of another 10,000 years long geophysical/radiative event between 84,000 and 74,000 years ago, and which event seems responsible for the first "art revolution" in South Africa. There is possible that this mentioned event had a specific influence at the beginning of language.

As I. Tattersall suggested, "*this was a configurative phase in language evolution*" that may follow up some other previous geomagnetic events. The complete externalization gradually occurred in an era that begins 42,000 years ago.

In the meantime, archaeological evidence points out that our braincase reached its current shape and expression mostly during Holocene, around 8,000 years ago.

As it appears, the globularization appeared (35,000 years ago) shortly after the geomagnetic, and cosmogenic/radiative events turned manifested while affecting our brain biology. Still, the process continued to modify brain architecture until at least 8,000 years ago.

Archaeological evidence points out that The Early European Modern Humans (EEMH) appeared in Europe starting 48,000 years ago and had a braincase distinct from 30,000 years ago.

The researchers describe such differences analyzed on archaeological skeletal remains. EEMH had a more robust body associated with: broader and shorter faces, more prominent brow ridges, more rectangular eye sockets, shorter upper jaws, horizontally oriented cheekbones, and more prominent teeth. However, any selection for lighter skin would not become prevalent before 30,000 years ago.

These mentioned characteristics of the EEMH's face configuration indicate causal changes in frontal lobes' size and other lateral brain areas' changes.

For example, a size increase of regions adjacent to the parietal area is associated with a parietal bulging. However, parietal bulging in present-day humans is not linked to significant variations in the shape of the precuneus that is the brain organization's hub. Precuneus is significantly larger in humans than in chimpanzees. It is related to cognition and specialization, which explains its significant but unique augmentation in the human brain.

It was found that precuneus has a paramount role in processing unconscious information, generating unconscious thought.

During most of the Paleolithic, unconscious thought was prevalent in Homo sapiens.

The change toward defined modernity is connected with the recent evolution toward bulging the thalamus found to be the central hub. The nonlinear information, aka unconscious, turned transformed into the linearity express by our conscious states. Also, the thalamus seems to indirectly govern the Boca and Wernicke area, which process the language. The same thalamus appears to process the initial nonlinearity and complexity of language while transforming it into its well-known linear expression.

As long as the language evolution is documented as an event occurring some 30,000 years ago, the thalamus bulging must have happened prior but not far distant from that revolution. Such a bulging significantly contributed to a language revolution, and since then, it has increased our conscious thought exponentially.

However, the simultaneous increase of precuneus shows that much of the conscious thought originates into the precuneus' nonlinear processing. Still, it is linearly reduced by own thalamus processing and distributed further to both of these language-specific areas.

As behavioral modernity features accumulated in the human brain, this process seems to reach a peak, and a significant marker in the era placed 50,000-40,000 years ago. This era overlaps the Laschamp geomagnetic excursion and associated magnetic pole reversal, which simultaneously occurred 42,000 years ago.

However, a clearly proven tendency toward skull globularity did not manifest before 35,000 years ago.

Niego and A. Benitez-Burraco, in a study on *Williams Syndrome* (published on March 18, 2019, in Frontiers), indicate that "*human language resulted from changes in our biology, behavior, and culture.* They found that *"William syndrome (WS) is a condition with a "clear genetic basis, resulting in a distinctive behavioral and cognitive profile, including enhanced sociability. In this paper, we show evidence that the WS phenotype can be satisfactorily constructed as a hyper-domesticated human phenotype, plausibly resulting from the effect of the WS hemideletion on selected candidates for domestication and neural crest (NC) function."*

Wilkins et al. (2004) showed:

"Domestication usually entails a constellation of distinctive traits (the domestication syndrome), which are mostly related to the brain/cognitive changes. Such changes seem to be important for the development of language and Globularity."

"It has been considered that different external stimuli triggered the domestication syndrome. The result refers to the hypofunction of the neural crest cells (NCCs) in the embryonic phase of development."

"It is known that most adaptive differences between plants and animals are due to gene regulation changes and not protein evolution. Genome-wide patterns of gene expression can evolve by a broader mechanism, where the non-coding variation (RNA) has a significant role." It can be assessed that directional selection on RNA altered the expression of energy-metabolism-genes during domestication or self-domestication.

Different environmental conditions were manifested as stresses, while the gene activation came in response to corresponding biological characteristics, which are types of gene-expression.

Antonio Benitez-Burraco, Constantina Theofanopoulos, and Cedric Boeckx, in a paper published on June 28, Volume 37, Issue 2, pages 265-278 in Springer Link, considered that *"the genetic changes we have claimed led to globularization are intimately related to those affected NCCs, fueling (or being the main rationale for) the emergence of the (self)-domestication syndrome in our species and ultimately, of full-fledged languages."*

"The very nature of cultural learning appears to be key in capturing all this grammatical paraphernalia (Hurford 2011; Kirby 2012; Smith and Kirby 2008; Thompson et al. 2016). But Thomas points out, a major problem facing any attempt to account for language structure through a cultural mechanism is that the process required by such a mechanism is only possible if we assume the existence of preconditions, which we may call the "cultural niche."

"Sensitivity to communicative intent, at the heart of our communicative/pragmatic competence, characterize domesticated species in general. A cursory look at the fossil record highlight changes in anatomically modern humans that closely resemble the ones attested in domesticated animals: reduced skeletal and cranial robusticity, changes in dentition, retention of juvenile characteristics, changes in temperament (compared to other living primates), reduction of sexual dimorphism, etc."

"All these features suggest that Domestication may have taken place as part of the emergence of AMHs. Although, of course, once in place, it may have intensified in our recent history."

"It is a well-established fact that domesticated animals possess a distinctive, unusual suite of heritable traits not seen in their wild progenitors."

"We made as well an extensive survey of the PubMed literature to know whether our candidate (genes) can be regarded as "neural crest genes." The whole list of them, currently encompassing 152 genes, shows that 60% of them play a role related to the neural crest." Here I suggest that the correlation of several elements that contributed to the development of language, and where one of them was self-domestication, have been events that occurred in the same period 35,000-10,000 years ago. It further led to another process: that of

Domestication of animals and plants, which seems to have started with dogs some 15,000 years ago.

Such correlations can also be drawn between cognition, culture, and art. The art explosion that dates some 38,000-30,000 years ago also overlaps the language explosion and corroborates very well with a manifestation of human self-domestication, followed by the Domestication of animals and plants. Here there is the origin of culture, too.

Benitez-Burraco and C. Boeckx published on June 16, 2015, Frontiers in Psychology, a paper named "*Possible Functional Links Among Brain-and Skull-Related Genes Selected in Modern Humans.*" The authors reveal: "*The genes we focus on are found mutated in different cognitive disorders affecting modern populations, and their products are involved in the skull and brain morphology, and neural connectivity. It means that the **changes affecting most of these proteins resulted in a more globular brain and ultimately brought about modern cognition**, with its characteristic generativity and capacity to form and exploit cross-modular concepts, properties most clearly manifested in language*".

The human brain has experienced an exceptionally high level of positive selection compared to the stomach or heart. For example, the NOTCH pathway works differently in human and macaque organoid-test tube models for the developing brain. David Haussler, a bioinformatician at the University of California, Santa Cruz, found that the NOTCH2NL was missing in the macaque organoid as in other nonhuman apes as well.

In a study published in the spring of 2018 in the publication The Cell, Vanderhaeghen, and his colleagues indicate that NOYCH2NL works to bust neuron formation.

As Ian Sample explains in his article in Neuroscience (published on February 26, 2015): "Researchers believe the gene plays a pivotal role in human cognition by ramping up dramatically the number of neurons in the cortex, a brain region that is central to reasoning, language, and sensory perception."

The gene known as ARHGAP11B is very highly active in stem cells that make some neocortex neurons. It generates a significant multiplication that contrasts with its activity in other animal brains.

It is also known that individuals that are younger than 35,000 years ago overlap with the range of variation in present-day humans. In my opinion, changes in brain morphology and connectivity changed the shape of the skull to globular and helped develop language skills. Modern behavior refers to creating tools and art, but secondary to language itself, representing communication, planning, control, and socialization. Scientists call such a period of their emergence the "human revolution" that started approximately 50,000 years ago.

However, taking this advent into more detail, most scientists agree to reduce this interval to the last 40,000 years, while the spherical features started to emerge only 40,000-35,000 years ago clearly.

In an article published in Science Advances in January 2018, Dr. Simon Neubauer of Max Plank Institute for Evolutionary Anthropology, in Leipzig, Germany, indicates:

"Globularity itself likely didn't give us advantages, but the features, that contributed to the rounding, probably did: the bulging of parietal areas and the bulging of the cerebellum. The parietal lobe is an important hub in connecting brain regions and is involved in functions like orientation, attention, and sensorimotor transformations that underlie planning and visuospatial integration. Meanwhile, the cerebellum relates to motor-related functions, like balance, and integral functions like working memory, language, affective processing, and social cognition. It's likely the emergence of these skills prompted the human revolution."

"Parietal areas are involved in orientation, attention, perception of stimuli, imagery, self-awareness, numerical processing, long-term memory, and tool use."

However, the thalamus is central to language and human cognition, as it modulates frontoparietal activity. A gene-set is where its members are involved in the development and connectivity of the thalamus in the brain and produce language-associations.

"It is also interesting to point out that, in present-day humans, brain globularity emerges developmentally during a few months around the time of birth that is a critical and vulnerable period."

"The globularity of modern man's endocranial cast is directly correlated with both hemispheres of the cerebrum, *and both hemispheres of cerebellum-which make up the entire cranial globe- becoming more like a spherical volleyball and less like an oval-shaped rugby ball."*

"In the first three months of life, the cerebellum grows at the highest rate among other brain parts. The characteristics of globularity seem to evolve during a prenatal and early postnatal period".

Jean-Jacques Hublin, from the same researcher team in Leipzig, said:

"The gradual evolution of modern brain shape seems to parallel the gradual emergence of behavioral modernity as seen from the archaeological record."

Importantly, these shape changes evolved independently of brain size.

There is known that the human brain entered a size-reduction process that begins to manifest around 20,000 years ago.

Cedric Boeckx and Antonio Benitez-Burraco, in another study (*"The Shape of The Human Language-Ready Brain"*) published on April 4, 2014, in Frontiers in Psychology, explain:

"The anatomical shift that led to Globularity also entailed significant changes at the subcortical level. Here we focus on the thalamus, which we argue is central to language and human cognition, as it modulates frontoparietal activity."

"It enables us to keep clearly separate two entities: one, the language-ready brain, understood as the cluster of brain properties that sets the stage for language ontogeny and phylogeny, and the other, language, understood as the collection of properties that humans eventually acquire as a result of social interactions."

"Our conclusion is also in line with more recent studies casting doubt on a direct link between laterality and language as a whole.

Thus, to the extent that laterality bears on the linguistic brain, we think that it is likely to be tied to the communicative function of language, or the externalization component".

"A detailed examination of endocasts from fossil specimens has revealed that modern humans, in contrast to otherwise heavily encephalized Neanderthals, show species-specific neomorphic hypertrophy of the parietal volume, leading to a dorsal growth and ventral flexion (convolutions) and consequent Globularity of the whole structure. Subsequent research has established that Globularity is the result of a unique developmental trajectory in modern humans, taking place at the stage of growth where the brain is the primary determinant of skull shape".

"Globularity is what underlies our species language-readiness."

"Globularity de-emphasizes the role of frontal lobes in giving rise to modern cognition. The literature on Globularity indicates that at the very least parietal volumes are equally important. Only Homo sapiens show a generalized enlargement of the entire parietal surface. It is indeed reasonable to think that the morphological changes in the

parietal region are to be related to important neurofunctional consequences, completing the functions of the frontal lobes".

"Globularity is not just a superficial property of braincases. It crucially entails modifications of neural connections. The developmental trajectory giving rise to Globularity is critical to forming a network of neural connections capable of supporting the most distinctive cognition mode that numerous scholars associate with language. Put succinctly, a globular brain gives rise to the language-ready brain".

So, the **frontoparietal-thalamic network emerges in the context of Globularity**. As it seems, the dorsal thalamus, specifically the pulvinar and the mediodorsal nucleus, played a very significant role. Here occurred a coordinated expansion of functionally and anatomically related areas, and in my hypothesis, such a complex process **resulted from a highly stimulated neurogenesis in the Dentate Gyrus**. Such a phenomenon occurred in the interval 40,000-10,000 years ago. It overlapped geophysical events' mentioned chain when the atmospheric concentration of the C14 isotope was the highest on a long geological scale (approximately 30,000 years). I like to mention some craniometrical studies' results, enhancing several other features produced by the skull's globularization.

Such studies indicate that reduced facial projection results from spheroid shortening and generates other phylogenetically and functionally significant differences in craniofacial shape.

In the following paragraphs, I will refer to the research and the observations made by Daniel E. Lieberman with the Department of Anthropology, Rutgers University, New Brunswick, and the New Jersey campuses.

Hence, *"reduced facial projection increases overall cranial Globularity by decreasing the cranial length relative to endocranial volume. It could explain how modern humans have evolved rapidly from more archaic forms. One possibility is that a shorter spheroid, by decreasing the oropharynx length, has been an adaptation for speech, contributing to the vocal tract's unique proportions. The horizontal component is equal in length to its vertical component. This is in contrast with a markedly longer, as it exists, in other primates. Such a configuration generates an ability to produce acoustically distinct speech sounds"*.

However, the same reduction in facial projection helped humans develop more significant facial gestures, which were initially used in sign languages and later in parallel with vocalization.

"Beyond the notion of involvement of the thalamus and the basal ganglia (BG) in linguistic operations, more specific functions of deeper structures remain controversial. In such a view, the thalamus does not engage in proper linguistic operations. It seems to act as a central monitor of language-specific cortical activities but supported by basal ganglia in all these activities".

Brain-imaging methods revealed that much more of our brain in language processing than previously thought. We now know that numerous regions in every major lobe (frontal, parietal, occipital, and temporal lobes), especially in Broca and Wernicke areas, connect with the cerebellum, are involved in the ability to comprehend and produce language.

Several authors have recently concluded that there exists a limited degree of spatial integration between soft and hard tissues, indicating that Globularity and the development of various brain regions have mostly been independent events.

An essential aspect refers to how the visual brain has switched to the verbal brain, and if here, the globularization would have played a distinctive role or not? Then, the researchers *"hypothesized that if the visual cortex was involved in language processing, those brain areas should show the same sensitivity to linguistic information as classic language areas such as Broca's and Wernicke's areas."*

The last question was formulated by Marina Bedny, an MIT postdoctoral associate in the Department of Brain and Cognitive Sciences. Her response paper appeared in the Proceedings of the National Academy of Sciences on February 28, 2011.

Her research found that "*the visual brain regions were sensitive to sentence structure and word meanings in the same way as classic language regions.*" She found that "*these brain regions can go from vision to language, indicating that the intrinsic function of a brain area is constrained only loosely.*"

Amir Amedi, a neurophysiologist at the Hebrew University of Jerusalem, shows that the left occipital cortex is processing language. He said: "*I think it suggests, in principle, and if the changes are forced early in development and early in life, any brain area can change its skin and do any task or function.*"

Bedny notes that "*it suggests that a part of the brain can participate in language processing without having evolved to do so.*" She also thinks that this case occurs "*due to a natural redistribution of tasks during brain development.*" "*As these brain functions are getting parceled out, the visual cortex isn't getting its typical function, which is to do vision.*"

In sum, her study indicates that "*different parts of the visual cortex get divvied up for different functions during development. A subset of* (left-brain) *visual areas appears to be involved in language, including the left primary visual cortex*".

Another paper by Sarah Genon, Andrew Reid, Robert Langner, Katim Amunts, and Simon B. Eickhoff was published in Trends in Cognitive Science in January 2018, under the title "*How to Characterize the Function of a Brain Region.*"

The authors show that "*a large-sample data can disclose covariation between brain region features and ecological, behavioral phenotyping.*"

"Current conceptualizations of brain function as a Bayesian machine, in which brain areas are seen as connected and relatively specialized computational units, are in contrast with actual available knowledge about functional specialization."

"We propose that assessing the relative functional specialization of brain regions requires a critical change in viewpoint, wherein a priori defined construct is the brain region and the unknown are the behavioral functions associated with it."

*"The cerebral cortex is not a homogeneous entity but can be subdivided into regionally distinct modules (cortical areas or subcortical nuclei) based on functional and structural properties. The latter emphasizes that no brain region is by itself sufficient to perform a particular cognitive, sensory, or motor function. Rather, **all mental capacities rely on a dynamic interplay and exchange of information between different regions**".*

"Accordingly, each area can perform a limited range of functions, but the concrete behavioral output depends on which other areas (based on efferent connectivity)."

The study refers to a type of <u>coactivity</u> (as information is sent and received) <u>of each region and between regions</u>.

These recent researches change our previous concepts about how the new verbal functions took over the old visual roles. In the meantime, it suggests that a particular increase in neurogenesis has taken place in the recent past, and it contributed to a significant increase in brain plasticity as it becomes manifested in various brain regions.

As it appears, the process of <u>overlapping of verbal functions on visual functions becomes possible by a significant increase in plasticity</u>, and it probably does not diminish or eradicate the role it overlaps, as it was previously thought.

In support of the **idea of coactivity,** I like to quote another paper by Antonio Benitez-Burraco and Elliot Murphy, published in Frontiers in Behavioral Neuroscience on August 22, 2019, under the title "*Why Brain Oscillations Are Improving Our Understanding of Language?*"

The authors revealed: "*It is now evident that <u>language results from the coordinated activity of several widespread brain networks</u>, encompassing different areas of both hemispheres. Oscillations enable the construction of coherently organized neural assemblies through establishing transitory temporal correlations. They reflect synchronized fluctuations in neuronal excitability and are grouped by frequency*".

"*For instance, they are primitive components of brain function and appear to be both domain-general (i.e., individual oscillations intervene in different cognitive and perceptual functions) and domain-specific (i.e., there exists a specific pattern of coupling between oscillations related to, and explaining each cognitive function).*"

"*Thousands of biological factors interact to regulate language development and processing. Also, brain oscillations are highly heritable traits, including oscillations related to language.*"

"*Several linking hypotheses, connecting particular genes and oscillatory behavior implicated in language processing, can be posited, suggesting that **much of the oscillome is likely genetically-directed**; the set of genes implicated here is termed the oscillogenome. Importantly, these candidate genes map on to specific aspects of brain function, particularly on to neurotransmitter function, and though dopaminergic, GABAergic and glutamatergic synapses*".

"*A system biology approach to language is preferable since it allows us to understand how <u>language emerges from the complex interactions among thousands of biological factors</u>, <u>most notably oscillations</u>. It is now clear that because language evolved mostly as a result of specific* *changes in the developmental path of the hominin brain in response to changes in the environment in which our ancestors lived, we need to consider developmental, evolutionary aspects on a par."*

"The oscillations can explicitly be linked, in some way, to all major topics in the study of the computational nature of language. Specifically, oscillations might be a better (or perhaps, the optimum) candidate for properly defining the morphospace of the adaptive landscape of language growth in the species".

All oscillations do is to put together neurons and groups of neurons by a coactivation mode.

* * *

However, when one watches a map of the language-related areas' distribution, one observes that all these areas surround the brain's core, where the thalamus is located. They seem interconnected with the thalamus and hippocampus.

Nevertheless, some philosophers debate the issue if we master the world by language, or it is the case that we only follow the language as it speaks to us.

Here, there is a debate about if the language is a representing, or as a propositional, or as an object. The answer to this debate is biased when it indicates that language would be an anatomical-technology or behave in such a technical manner.

For example, the Buddhists do not seek an experience that would transcend language, but they seek to free the language of its artificial boundaries. Those boundaries appear in our metaphysical thinking that sees the language as a tool for our use.

* * *

Kate M. Lesciotto and Joan T. Richtsmeier, in their article *Craniofacial skeletal response to encephalization*, published in American Journal of Physical Anthropology, Volume 168, Issue S67, on January 24, 2019, explains:

"*Another potential explanation for the globular shape of the human cranial vault has been proposed as part of a **"wiring" hypothesis, wherein the shape of the human brain is attributed to the need to reduce the wiring length** (distance between the axons and neurons to form functional circuits), both within the telencephalon and between the cerebrum and diencephalon (e.g., Bruner, Martin-Loeches & Colom, 2010; Chklovskii & Stevens, 2000; Hofer, 1969; Lieberman, Ross, & Ravosa, 2000: Mitchison, 1991; Ross & Henneberg, 1995, Sporns, Chialvo, Kaiser, & Hilgetag, 2004; Van Essen, 1997). This idea can be traced to Santiago Ramon y Cajal and is not specific to the explanation of human brains, as applied across species. (Rivera-Alba et al., 2011; Stevens, 2012).*"

"Some scholars have hypothesized that **a more spherical cerebrum would optimize neural connectivity of axons and dendrites within the evolutionary expanded hominin neocortex** (Bruner, 2004; Lieberman, Ross, & Ravosa, 2000; Ross & Henneberg, 1995). The cerebellum and brain stem location would anatomically prevent the neocortex's expansion in a posterior or inferior direction, making anterior expansion the path of least resistance. To produce a more balanced spherical shape alongside the anterior neocortical expansion, the brain necessarily develops a "kink" or ventral brain flexion (Lieberman, Ross, & Ravosa, 2000)."

"At the time when brain globularity fell within the range of variation of present-day humans, the full set of behavioral modernity had been accumulated around 40,000 years ago."

I like to hypothesize that there had been a connection between language development, especially manifested as the language revolution, and the changes in the brain's anatomic architecture.

Definitely, the language brought to the brain a lot more information to be processed, and such a stimulus produced a neural adaptation in the brain network of circuits. Eventually, the adaptation came in the form of shorter circuits around a central core of processing. The adaptation was facilitated by a stimulus applied to that neurogenesis produced by the Dentate Gyrus, also placed in a prominent brain position. As aforementioned, the stimulus increased the atmospheric concentration of the C14 isotope that started 42,000 years ago. The result of the geomagnetic/radiative stimulus was a significant increase in the brain plasticity that generated a new architecture where the circuits turned shorter. Hence, they became more efficient in dealing with that information produced by the rapid language evolution. In my opinion, the change in architecture should be on par with the evolution of language. When we combine these two features or processes, the apparent result is brain globularization.

"Language revolution," assumed to occur somewhere between 35,000 and 25,000 years ago, had produced a need for more energy: the processing of such language-related information requires more energy consumption.

As the archaeological skeletal investigation attests, another process developed in the last 20,000 years ago indicates a gradual reduction in human brain size.

However, the skull's globularization does not correlate with the brain-size, meaning that the globularization process does not depend on the brain's size.

As it can be assumed, the globularization reduced the connecting distance between the neural circuits, and probably it led, in the first stage, to a diminishing of energetic consumption.

In the meantime, globularization manifested by expanding several brain regions, like the parietal lobes and some centrally placed areas. These expansions meant new gray matter and new circuits, which altogether increased the brain's processing capabilities.

The highest energetic consumption seems to be produced by the activity of memory. Storing memories increased as the language and social relationships evolved.

As I mentioned before, the language came within its evolving complexity, where the attributes of planning, control, socialization came together with the utterance. Thus, this complexity evolved in all its components.

The complexification of language continued, and the initial small "saving" turned insufficient for a continual increase in neural energy consumption.

Overall, prehistoric progress was significantly small compared to the language attributes existing around 2,000 years ago.

One can assume that this small "saving," obtained by shortening the circuits' length, was exhausted around 20,000 years ago.

And then, another process started: a diminishing in the size of the brain. All humans did was to learn. Hence, learning and storing daily experiences became vast consumers of the brain's energy.

How could this new process of brain-size diminishing be explained?

Obviously, it must be connected to the consumption of energy.

Here were two ways to approach: one referred to a possibility to make the neurons learn slower, saving a lot of processing energy; the other way was offered by verbal communication in itself because the visual perception (that is a process of opticoding by opticoders) is much faster (probably at least 10,000 times) that processing of verbal perception (lexicoders).

So, in the first place, the **language's increased information was balanced by its slower processing** within the neural network. For a while, this evolution was on par. The most notable alternative to increase the amount of energy provided to the brain is to improve the diet.

However, various research that compared the Paleolithic diet to the current diet concluded that both are almost equal in energy production.

Thus, the introduction of agriculture did not change in a significant manner this energetic balance.

It remains to disclose how the neurons have turned to low energy consumption when confronted with learning tasks and memory storage. The memory storage procedure shows that one has to erase something old or less critical when storing something new. Thus, the energy needed to hold something new can be recovered by erasing an old memory. Here the balance is preserved. Nevertheless, this process does not influence brain-size.

Now we came to the last element of this discussion: the slow-energy consumption of the neuron itself. As it seems, such stringency occurred in time starting 20,000 years ago and accelerated ever since until 8,000-6,000 years ago.

Learning is a transformation of neural networks at the expense of energy.

The above aspect was the article's subject (*Stochastic Thermodynamics of Learning*) and Sebastian Goldt and Udo Seifert's research and published on November 28, 2016 Xiv:1611.09428v1. The authors suggest that the thermodynamic cost of learning bounds information acquired by the neural network. The organisms (bacterium in the initial case study) exploit a given energy budget to adapt to the environment. Stochastic thermodynamics studies the interplay of information processing and dissipation in interacting, fluctuating systems far from equilibrium.

"*Learning is about extracting models from sensory data. It is implemented in neural networks in living systems where vast numbers of neurons communicate via action potentials. The electric pulse is used universally as the basic token of communication in neural systems. Action potentials are transmitted via synapses, and their strength determines whether an incoming signal will make the receiving neuron trigger an action potential of its own*".

The authors indicated that learning efficiency is constrained by the total entropy production of a neural network. They noticed that the <u>slower a</u>

neuron learns, the less heat and entropy it produces, which increases its efficiency.

This study, in my opinion, suggests that there is the possibility that the human brain constrained by a high entropy produced by the language was forced to adapt by changing the efficiency of its neurons. The adaptation, as mentioned above, was favored by a period of 35,000 years (from 42,000 to 5,000 years ago) when the concentration of atmospheric C14 intermittently was 60% to 100% higher than the current one. As a result of this high concentration, the Dentate Gyrus, in particular, was stimulated to increase its processes of neurogenesis significantly.

During the said period, an increase in the need for more metabolic energy to the brain occurred. Hence, as neurogenesis products, the new neurons came more ready to adapt to the imposed energetic task. I suggest that these new neurons suffered the needed alteration that made them slow-learners. In this manner, the new neurons consumed less energy for processing the complexity of language.

However, new neurons' efficiency has led to the cancelation of those other neurons, which were no more needed. Probably, in time, the new neurons but highly efficient ones came in smaller numbers, causing a volume reduction.

This **volume loss refers to gray matter that diminished the brain volume as a whole. Previously, the shortened circuits have produced a reduction in the volume of white matter**. There were two distinct processes, where both contributed to the general shrinking of the brain volume.

Consequently, the skull adapted, shrinking in the same proportion. There are many details of this process, which the research must establish exactly.

The globularization continued to evolve in the sense that it gradually affected deeper brain circuits, making them more efficient, shortened their length, and in the process, some more white matter (representing those circuits) became unnecessary due to shortcuts, and the neurogenesis did not renew it.

In my opinion, increased neurogenesis (that would refer to the quantity and the quality of those new neurons) had an essential contribution to all processes mentioned in this paper, being that fundamental factor that allowed such significant changes in our brain architecture. Once again, our modernity started 42,000 years ago and evolved gradually to its current status, being uniquely intermediated by an era of high concentration of C14 atmospheric isotope that directly influenced brain metabolism and neurogenesis.

Fourth Paper
Changes in brain lateralization

Conjugates Switched the Brain Lateralization and Generated A New Type of Intelligence

A conjugated reaction refers to neurons' temporal spikes due to increased energy provided to the brain by an increased entropy.
The research results from various experiments suggest that increased stimulus generates an increase in inhibition.

However, blocking-inhibition in random neural networks reduces response variability and increases activity sensitivity within the stimulus's location.

In this case, the inhibition acts as a dispenser and sharpens the stimulus contrast (Netta Haroush & Shimon Marom, March 19, 2019, in Scientific Reports 9, Article number 4969-2019).

In increased entropy, the stimulus mentioned above acts to disperse the inhibition while enhancing the stimulus only locally.

In simple words, the increased entropy is increasingly filtered by the inhibition; thus, the neural activity enlarges, and an increased amount of information is provided to the conscious domain. It augments consciousness.

However, the brain hemispheres have developed a type of conjugated activity: the right hemisphere is characterized by hyperactivities, while the left hemisphere shows hypoactivities, which are inhibitory in nature.

For example, the right hand's control is driven by the left hemisphere; this hand is the most used because the inhibition processes assure the most accurate control of the right-hand movement. Prehistoric people were hunting and building their tools with an overwhelming contribution of their right hand.

The *right hemisphere* locates the visual activity that *implies a lot of sensitivity*. It is known to be the seat for music and rhythmic acoustics. It was the dominant hemisphere that generated the visual-thinking. Thus, its inherent sensitivity allowed the development of emotional thinking.

Not coincidentally, the sign language developed as a first because of the right dominant hemisphere's role in guiding the signing's visual element. Hence, signing evolved from visual thinking.

The research has determined that lip-smacking (as a type of sign language), as observed in several primates, occurs in the same theta rhythm range as human speech. Both sign and spoken languages seem based on a pre-existing perceptual neural oscillation, like the theta rhythm.

However, the control of a more complex signing needed the control and planning of the neural capabilities mainly developed in the left hemisphere.

The evolution of language needed a better and more complex modality to communicate, and as their anatomy evolved, humans opted to refine the acoustic component.

At that time, the sign-language evolved by inserting grammar rules, which rules came from that hemisphere with higher capabilities to control and plan, which was the left hemisphere.

The acoustic component of language needed an even higher type of control (for modulation and grammar), and therefore, the only option available inflicted the full contribution of the left hemisphere.

As the language evolution turned into a success, by better communication, improved planning, control, an increased rate of socialization, and domestication/self-domestication, the left hemisphere's role became increasingly prominent.

At some point in time (assumable some 4,000 years ago), the hemispheric dominance switched utterly, making the left hemisphere the ruler of the brain.

I used this short review of the *visual conjugated relationship with the evolving verbal mode* to suggest that **our brain evolution was based on a conjugated relationship between the brain hemispheres**.

Even when brain lateralization is similar in all animals, our prehistoric ancestors succeeded in using a narrow niche that contributes to developing a novel type of communication among mammals. It was the effect produced by the sensitivity to the initial conditions.

The switching in hemispheric dominance seems to be a novelty in the animal world that generated an anatomic type of technology (language). Other beings use anatomical technology, but the complexity developed in humans exceeds everything that is currently known.

Maybe the most significant effect of such switching is reflected in "intelligence." Human visual nonlinear thinking changed into verbal linear thinking, and **right brain intelligence turned into left-brain intelligence**.

Here, I have to say that **these two types of "intelligence" evolved in a conjugated relationship** while being a direct result of the effect (while temporal) of the C14 isotope on the DNA architecture of the Dentate Gyrus in the hippocampus and affected the very process of neurogenesis.

The high concentration of the atmospheric C14 isotope repeatedly occurred in the last 42,000 years. It constantly but temporarily affected the Dentate Gyrus architecture, suggesting several corresponding neural mutations or significant neural gene expression changes.

In sum, after several transitional stages, starting 42,000 years ago, an ever-accelerating process of dramatical changes occurred between 10,000 years ago and 4,000 years ago. I would presuppose that about 4,000 years ago, the new architecture of the human brain was already established, and the new intelligence became effective.

As Cheri Florance, Ph.D., explains, "the eye-to-brain pathway is called *opticoders* (the visual brain), and the ear-to-brain pathway is called the *lexicoders* (the auditory-verbal brain). Even today, 32% of the population is born with a visual brain. This means that these people use their *opticoders* to think and process information". "The *opticoders* brain is considered to be 20,000 times faster than the lexicoders brain, leading to the type of visual, outside-the-box thinking that has resulted in some of the world's greatest inventions and creative contributions".

The above paragraph indicates the difference between the old visual brain and the new verbal brain, as it was estimated in our time.

However, many scientists see the old thinking mode as dominated by a type of *correlative thinking*. The relative difference between correlative thinking and logical analysis means that the *ambiguity, vagueness, and incoherence associable with images and metaphors, which are specific for correlative*, are carried over into the more formal elements of thought that describe the logic. Correlative thinking involves the association of significances into clustered images, which define the context. **Correlative is a contextual "intelligence."**

One of the critical devices for resolving the ancient correlation was the contrast of *yin* (shady side) and *yang (sunny side). Correlative descriptions are prescriptions.*

Since the old emotional thinking began losing ground against logical thinking, the decrease in emotional began to elevate the power of concentration that increases the *content-driven analysis* as a new intelligent behavior (as expressed in today's IQ scores).

There is a saying: you cannot be rational if you are too emotional, but you cannot be rational if you are not emotional. Your emotional brain represents your instincts, impulses, and intuition. Rationality represents some 20% of human decision-making. Emotions still drive 80% of our choices.

Current statistics indicate the 25-30% of people use visuals as the primary mode of activity. 25-30 % use verbal as the main activity mode, while the remaining 40-50% use both modes. This information discloses that visual thinking is today, only 50% of total thinking. And the remaining 50% of our brains' verbal activity occurs at a mental speed that is 20,000 times diminished compared to visual speed.

Overall, at least half of our brain activity occurs at dramatically reduced speed (approximatively 20,000 times slower). This is our new verbal brain.

A team of researchers led by Dr. Sebastian Goldt (University of Stuttgart, Germany) has discovered that "the slower the neuron learns, it increases the neuron efficiency." The study targeted a recent adaptation of the neurons in the human brain to use less energy for their functioning. Slower learning diminishes energy consumption.

This adaptation came in response to changes in the human diet (that seem to provide less energy to the brain), reduced physical activity, and a significant slowdown in the verbal neural networks' processing speed.

However, some recent studies suggest that the hunter-gatherer's daily energy requirements were likely no different than those for current Western populations. The adoption of agriculture increased Neolithic foragers' workload because the older foragers enjoyed spending a moderate amount of time on subsistence work each day. Thus, ancient foragers consumed less energy for foraging than the farmers for land cultivation and domestic work.

The foragers provided more energy to their larger brains, which brains employed almost in exclusivity their visual but high-consuming-energy circuits.

As the language evolved, the verbal but slower circuits interfered with and/or overlapped some of the visual circuits (statistically, about 50% of them). The oral mode consumes less energy. But the introduction of farming increased body energy consumption, providing less power to the brain's needs. At this moment, the brain's energetic balance was affected and probably forced to operate a neural adaptation to lesser energy provided to the brain.

It might explain why the skull's size decreased more than 25% in the following few thousand years: gradually, the brain workload diminished by 10%-20%-30%-50% because of a corresponding reduction in the neural working speed. Slower speed implies less mental work.

The verbal mode needs for neural circuits probably resumed to invade and take over many existing visual circuits that gradually turned relatively less active instead of building up their own verbal circuits.

As the process above continued, the development of the conjugated relationship in corresponding pairs continued, too. It has contributed to a significant increase in the left hemisphere's overall activity, consolidating its dominant role.

As the left hemisphere activity increased, human language evolved, logic thinking was introduced, mathematics was rationally configured, imaginary-thinking flourished, thinking concentration increased, emotional thinking rapidly decreased.

<u>The slower processing of perceptual information made humans see more details of reality and increased their power to discover.</u> The perception of time gradually changed, too.

I assumed here that several neuronal mutations have occurred in the last 42,000 years. But the genetics research seems to do not find any. This aspect may imply that all forcing factors, which intervened in this time interval, have caused only a genetic new expression.

However, as I assumed, <u>everything that happened was driven by repeated stages of conjugation between the brain hemispheres</u>.

It is like having a house with only two rooms, and in each room, there are certain things. At some point, one started to move stuff from a crowded room into another but less crowded space. When this action ends, we still have the same house with two rooms, with overall absolutely same things, but each room's stuff shows a changed arrangement.

In sum, for an outside observer, absolutely nothing happened in this house because no new things were added or removed from it.

It probably is the case of our brain. We still have the same Paleolithic brain (because no significant mutations occurred), while the neural circuits seem differently interconnected, favoring the connection between closer circuits. Some brain closer sections turn very active, while the distantly placed cells lose much of the previous interconnectivity.

Then, what happened in this house? An outside factor increased the plasticity (or movability) of almost all things placed inside the house, and this plasticity/movability allowed the staff to be easily moved around.

Inside this house, the staff is absolutely the same. Things now have a new functionality resulting from a new type of interconnectedness because of the arrangement changes.

Is that all? Not really. The changes inside the house have a significant influence on the individual living here. In what sense?

The inhabitant becomes inspired to change the perspective on the practice performed outside because it becomes possible to plan, interconnect, and re-arrange in novel ways such as work, predict, and obtain significant new functional benefits. An "imaginary-thinking now inspires everyth*ing*."

Even then, the power of imagination does not help know how the outside world is made. But, he/she could invent and construct an outside world that works only for itself.

In the end, one may have to confront the question: what would imply for our brain if humanity would manage to (somehow) transform the verbal language into an exclusively visual (optic) language?

The answer may indicate that our brain would have to use high-energy consumption circuits, increasing overall brain consumption over its peak in the Late Paleolithic. Now we add the complexity of language. The brain volume would significantly increase. We would need a much bigger skull (for a larger brain) than the largest skull that existed during the Paleolithic. And the neurons would have to re-adapt to higher speed processing. We would solve the problems much faster than we do now.

Such a future possibility of a visual (optic) language cannot be accommodated by our anatomy, except maybe in a very long-time interval of thousands of years, while it may imply the use of conjugates, which may not exist or not be available.

For the moment, we are constrained to rely on adding an exterior computing and problem-solving capacity, and Artificial Intelligence currently envisages that. This situation is generated by the fact that we are exponentially confronted with an ever-rising rate of artificial data produced by our civilization, which cannot be resolved by mental computation and investigation.

Here, there is no doubt that we tend to add an artificially created capacity to our little changed but Late Paleolithic brain. It does not matter how advanced this capacity would be because it would remain outside our brain and way outside our biological thinking capabilities.

Even the mental-interference of a proposed interface-approach currently developed by the Artificial Intelligence program cannot supplement the existing neural circuits in a truly biological way. Artificial Intelligence is not yet based on conjugated biological possibilities, which are naturally drawn to maneuver and accommodate our brain functioning.

In my opinion, for a very long time to come, the currently proposed Artificial Intelligence would remain separated from our brain biology and alien to ourselves.

The current situation may have to prioritize the research to investigate what other natural conjugates would be available for our brain system and mechanism, how they could be stimulated their occurrence, and explore what effects may generate such novel conjugates?

It may be the case that, in this manner, a lot of new but natural avenues can be unveiled, which may solve our problems outside of, while competing with the current Artificial Intelligence approach. Ultimately, they may help improve our sustainability, potentially resolving many severe environmental aspects.

It must be a new while a multidisciplinary field of competence.

As Helena Knyazeva says in her article *"The Complex Nonlinear Thinking"* (2004): "All systems of human knowledge are based on certain principles. The principles of complex nonlinear thinking bear in themselves the imprint of nature of principles as such. Thus, the **principles of thinking should be holistic, complex, and human-oriented**. The theory of self-organization of complex systems plays a peculiar role in the indispensable reform of thinking".

Undoubtedly, the aforementioned hemispheric switching represents an act of self-organization incited by a forcing factor and a particular stimulus. It has produced a switch in brain behavior and functioning.

In sum, the process of switching the conjugates sometimes started during the revolution in language that occurred at least 30,000 years ago.

The switching process gradually evolved during the entire period from 30,000 years ago until 5,000 years ago. This process produced the modern brain.

Fifth Paper
Universal technology of language and the intelligence of complexification

Universal technology of language

Abstract

This paper aims to demonstrate that in nature exist several fundamental mechanisms, which regulate the complexities. Such a regulatory process introduces two phases of complex systems evolution: the adaptability of undifferentiation and the differentiation's efficiency that causes stability.

The same two steps regulate neurogenesis as a complex system by producing undifferentiation and differentiation. It transforms multidimensional unstructured information into a lower dimensionality structured data.

This paper discusses why neurogenesis manifests distinctly in humans versus other animals and how this aspect makes an immense difference in achievement.

Further on, I demonstrate that the mechanism of codimensionality, covariation, and correlation generates a language-ready brain in humans. It serves as a mental blueprint that can assemble languages and also factual technologies.

As it appears, the language-ready brain was the result of anatomical and functional changes, which stimulated some processes of complexification where emergent behaviors arisen and produced intelligent behaviors.

The language-ready brain **developed in-depth** the processing of **linearity that is specific to all life forms**. Human language is initially processed in a universal mental mapping. Such nonlinear data is naturally reduced to linearity by applying selection, categorization, or segmentation that decreases the initial multidimensionality. **Every diversification produces a loss of the initial data.**

During most of the Paleolithic era, the brain used a prevalence of nonlinearly structured data (like olfactive, sensory, and emotional outcomes). The language development **highlighted the linear transformation mechanism** that increased **the diversification process** while **significantly extending the consciousness**.

Introduction

Humans share with other species the ability to produce a mapping containing strings of sound and meanings. But only humans can combine such lexical representations in novel ways. When one says combine, in this case, it has to imply a process of assembling-disassembling-reassembling. It is called syntactic processing that manipulates words and meanings, assembling them in sentences and other language formats. It is thought that behind this process is a type of biological machine that does assemble, combining words into larger structures.

Hence, *assembling* is the first part while the paramount one of this process. Here complex systems science describes a *complex system* as one with many components. They could be physical, biological, and social. The complexity characteristics cross unchanged between them. It is universal.

How a complex system gets so many components? It combines many simple elements. It assembles them as wholes.

What they have in common, while what distinguishes gases, liquids, and solids? It is an ***emergent*** behavior that arises from the components' free interaction, but the *emergent* aspect cannot be found in any of such components. For this reason, emergent behavior is defined as the intelligence of that system.

All complex systems display *emergent/intelligent* behaviors like general properties of those systems as wholes.

Why human brain came to this unique ability in the animal world to combine/assemble sound, words, and meanings?

Up to current knowledge, the best guess about how the language-ready brain was formed indicates that a significant change in human brain anatomy and neural circuitry favored novel complex processes within the neural networks.

A review by Blumstein (2009) says: *"The functional properties of language, that is, speech, lexical processing, and syntactic processing, appear not to be focally represented in one area of the brain; rather, each recruits a broadly distributed neural network of processing stream. Moreover, certain areas of the brain that have been associated with language processing appear to be recruited across cognitive domains, suggesting that while language may be functionally special, it draws on at least some neural mechanisms and computational properties shared across other cognitive and even visual domains."*

Another study by Evelina Fedorenko, Alfonso Nieto-Castanon, and Nancy Kanwisher with McGovern Institute of Brain Research, MIT (Published by Brian Lang in 2011 Feb: 120(2); 187-207) came to a similar conclusion:

"To summarize, no brain region has so far been convincingly demonstrated to be selectively engaged in syntactic processing or a component thereof. This may be that such regions do not exist. All regions that support syntactic processing support other, linguistic or non-linguistic functions."

The Science of Complexity better explains the situation described above: the neurogenesis pattern changes helped the human brain reorganize its network to allow all components to become interconnected in a complexity responsible for producing and developing the language.

This <u>same complexity generates *emergent* while novel behaviors that brought us the "intelligence"</u> paralleling the communication aspect. As a consequence, this novel "*intelligence*" helped to organize and to evolve the language. It also introduced a human capacity to assemble/disassemble/reassemble things exterior to us, and while complexifying them, we insert their own "intelligence" into them. This is an **"exterior intelligence"** added to our newly evolved **"mental intelligence."**

The scale of our mental complexities is the size of our "*intelligent*" behavior.

Let's see another aspect brought to us by the development of language.

According to Wikipedia, in 1975, Charles Berger and Richard Calabrese created uncertainty reduction theory "*to explain* **how communication is used to reduce uncertainties** *between strangers engaging in their first conversation together*." Claude E. Shannon and Warren Weaver (the fathers of information theory) suggested that "*uncertainty is reduced when the number of alternatives is limited, and the alternatives chosen tend to be repetitive*." This type of **selection increases predictability**. The horizon of predictability is often time.

Because technology follows the blueprint of language, such *technology is intended to reduce uncertainty, too*.

The structured data represents a mental **selection** that implies a quantitative definition.

Such definition can occur only when two or more elements can be **compared** and when **one element** could be **chosen as the reference term or the base of comparison. Comparison implies** a rapport or **a proportionality** that can be established between two or more entities. All measurements are the multiplication or proportional divisions of an element that is chosen as the base unit.

In language, diversification comes in the form of letters, words, numbers, etc., which are the symbols' quantitative transformation.

Prediction refers to a future state of a system. Its role is **to estimate how a motion follows commands**. Such a **motor prediction** is internal to Central Nervous System (CNS). It is not a fixed mechanism, and hence, as a plastic feature, it needs to be learned and updated by experiencing an error-and-trial procedure. It is common to all animals or beings.

The selection mechanism **is involved in prediction** because here, too, **it intervenes in a comparison between probabilities**, like a previously predicted outcome is compared with an actual one.

The **role of comparison** is to **reduce the uncertainty because it filters information, eliminating the unwanted or low-relevance data** that cannot be appropriately compared. It can also **highlight the data needed for prediction and control** or enhance other relevant sensory information.

It may be the case that comparison also relays on natural bridges of codimensionality and covariation. We can compare things only if they are somehow related, or they have something in common, like a "bridge" acts upon.

Structuring implies a **dimensional reduction**. Evolution means variation, and variation implies information. Selection chooses only the elements, which can adapt or fit changing conditions.

*The dimensional space reduction can occur only when two spaces of different dimensionality share one or more properties. I call the object of such sharing the presence of a **common factor** that is the property that is shared.* Such a "common factor" ranges from **codimensionality** to **covariation**, to **correlation**. A "common factor" is, in fact, a natural **bridge**. Covariance is maximal when the bond is linear. Correlation is scaled covariance.

A higher dimensionality translates into a lower one by operating a mental hyperplane processing that **relies on existing codimensionality** (we will see later the details of that process). Further on, the sharing or relationship is driven by **covariation** that exists between variables. When these two steps are accomplished, it may give rise to **correlations** developed at a linear level.

Thus, all quantities are arbitrary by their nature, and the relation between them is established by natural "**bridges**," which allow their computing when linearity is achieved.

For example, Chris Mitchell and Scott Nash (University of New South Wales) and Geoffrey Hall (University of York), in their study *The Intermixed-Blocked Effect in Human Perceptual Learning is Not the Consequence of Trial Spacing*, published in Journal of Experimental Psychology: Learning, Memory, and Cognition 2008. Vol. 34, No1, 237-242, give a significant explanation of the process of learning and differentiation of elements.

They show that "*intermixed preexposure, first tested in experiments on animals, in which the stimuli are presented in separated block trials (e.g., AX AX...BX BX), indicate a process in opposition to associative learning.*"

In contrast to the above "*separated block trials*," "*when two similar stimuli described as AX and BX are presented, they share a common X feature. This exposure of AX and BX appears to increase the effectiveness of A and B at the expense of X, rendering the two stimuli more different.*"

Now I will give my interpretation of such a process. My view indicates that the common factor described above (X) represents a codimensionality that is X. Here, AX is a larger space, and BX is the subspace. While dimensional reduction occurs, it projects the AX part (existing in a nonlinear space) into the BX subspace. As a result, the common factor X's importance or meaning is diminished so much until it is eliminated. The result unveils only the A and B elements, where A lost its original AX information to be incorporated into the BX subspace. Still, here, this subspace's characteristic is altered because the common factor X is removed from this subspace meaning. Ultimately, we perceive this phenomenon as a differentiation process, but where a significant information loss occurs.

It shows that the animals can compare the stimuli during preexposure and extract the unique features, making an apparent differentiation. Here, the differentiation allows comparing those stimuli that have a common feature (codimensionality or covariation).

It is experimentally proved that such a process of *intermixed blocked effect* occurs in human detection of flavor, in visual discrimination, but notably in verbal learning, expressed as *contextual interference*.

I exemplified how unconscious thought is projected into the conscious level by a linear in the above paragraphs. It also gives a glimpse into how linearity is segregated from nonlinearity.

Discussion

I will continue to discuss the unconscious vs. conscious because it refers to nonlinear vs. linear. Here, the nonlinearity of unconscious is referred to in scientific literature as "*holistic*," while the linear is described as "*seriality*."

I will quote a study of the Japanese researchers Tetsuya Kageyama, Kelssy Hitomi dos Santos Kawata, Ryuta Kawashima, and Motoaki Sugiura (*Performance and Material-Dependent Holistic Representation of Unconscious Thought: A Functional Magnetic Resonance Imaging Study*) published in Frontiers in Human Neuroscience on December 6, 2019.

This study suggests that "*we can think unconsciously. Unconscious thought (UT) refers to cognitive or effective decision-related processes that occur beyond conscious awareness, i.e., while people are consciously occupied in performing some other task."*

"Participants who are distracted show better decision-making performance compared with those who make an immediate decision (ID) and are allowed to think regarding their choices consciously.

A holistic representation of decision information is a multimodal, value-related cognitive process that may be associated with larger UT effects. It results in improved decision-making; i.e., holistic processing produces a large UT effect."

The study *"findings revealed a greater involvement of both the right dorsolateral prefrontal cortex (DLPFC) and middle occipital gyrus during UT. The occipital gyrus is considered to be involved in visual processing. Research has indicated neural activation of multimodal cortices such as medial prefrontal cortices and the precuneus, relevant to the holistic representation. Studies reported neural activation of the ventromedial prefrontal cortex, ventral Striatum, and posterior cingulate cortex. The ventromedial prefrontal cortex plays a role in subjective and emotional values, whereas ventral Striatum and posterior cingulate cortex are involved in reward value."*

"The cortical middleline structure, including the precuneus, is active during multimodal, which is consistent with the features of holistic presentation."

"The precuneus integrates multimodal information collected from various brain regions, thereby playing an important role in mental image processing."

"By contrast, decision-making from information serially presented makes participants process each alternative, significantly increasing the cognitive load. It consequently increases the processing, while it obstructs a comprehensive view over the complete information to consider. Here, again, in my view, the process of serially processing

discloses the issue of "*information loss that impedes the decision-making.*" The study found that a top-down decision is involved in the serial presentation.

"*Neural activation of UT is also observed in other decision-making scenarios, such as lie detection, responses to requests, moral situations, and product satisfaction.*"

I quoted the mentioned Japanese study extensively because it importantly discloses the UT's ability to learn unconsciously various tasks.

The above discussion may also play a significant role in understanding our prehistoric abilities before developing the language.

According to Wikipedia, "***pulvinar nuclei in the thalamus function***, *as a gatekeeper, is deciding which information should be inhibited, and which should be sent to further cortical areas. The CNS (Central Nervous System), after pulvinar nuclei deem the information to be relevant, acts as an essential inhibitory mechanism that prevents the information from flowing into higher cortical centers.*"

Hence, I would say that **conscious thought is shaped in the thalamus and must be related to language origin. Thalamus makes our world linear and serial.**

But the thalamus is not the only filter of reality. It is known that **Striatum** also plays a vital role in filtering information. As the basal input for the basal ganglia, Striatum receives a bulk of its incoming fibers from the cerebral cortex, modifies the signal, and sends it back to the cortex to generate a desired action.

It selects, as a filter, from a mixed nonlinear/linear outcome to the most desirable linear outcome. In this way, Striatum helps achieve a goal by selecting the appropriate action to be performed, but also it picks linearity out of nonlinear/linear mixed signal.

Another fundamental role of the Striatum is to produce a particular differentiation that <u>causes an inverse correlation</u> by which it identifies the linear outcome.

In modern psychophysical methodology, this property to generate **inverse correlation** provides proxies of the mental representation content that cannot be directly observed. For example, such a method can guide the decision-making on the path that produces a hidden reward. It finds such unseen rewards and directs the action to fulfill such a **rewarding scope**.

On the other hand, Striatum is seen as the mental center that facilitates addiction.

Here, the unseen is embedded into the nonlinear, and <u>Striatum can convert the nonlinear signal into a linear one</u>.

Here, there is no doubt that it employed the mentioned mental hyperplane by which it has reduced the nonlinear dimensionality to the linear one. It directly <u>participated in producing linear thinking by prioritizing linear goals</u> from a complex/nonlinear signal.

There is a scientific opinion that Striatum had a <u>particular contribution in gradually switching the right-hemisphere dominance toward the left</u>. It intervened in several ways in the competition posed by language development to the previous visual and sensory interpretation of reality. However, indirectly, it contributed to the setup of language in the left hemisphere.

In a New Scientist study, Emma Young estimated that unconscious processes at least 95% of brain information. Other scientists tend to allocate to it close to 99%.

Unconscious seems to have several types of filters, all of them based on various durations before a change can occur. But, all of them are changes occurring in less than 50 milliseconds. Here would be memories of less than a few seconds, which would not be perceived, but they have a specific while unknown mental processing role.

The Information Theory specialists estimate that consciousness selects a piece of information out of a million of them. Most (95% or even more) of that unselected information is processed at the unconscious level.

Recently, scientists turned to estimate that the human learning process can be understood as information compression. Similar principles apply throughout the nervous system and in much of the animal world.

Thus, as an intrinsic part of conscious processing, language deals with less than one in a million of the perceived information. But here, the development of language has produced a language-related while colossal multiplication of the initial data. Such new immensity is our new world created by language, and it is essentially imaginary while fundamentally distinct from natural reality.

When our brain architecture encountered a radical anatomical change (in my view some 40,000-30,000 years ago), very much like the changes dealt with enhancing the functional thalamus activity that directly stimulated the Broca and Wernicke linguistic areas of the brain.

Wikipedia says: "*The frontal lobe is involved in reasoning, motor control, emotion, and language. It contains the motor cortex that is involved in planning and coordinating movement. The prefrontal cortex is responsible for higher-level cognitive functioning, while it contains Broca's area that is essential for language production.*"

Here, I like to quote a seminal paper by Richard Passingham (*How good is the macaque monkey model of the human brain?* published in *Current opinion in neurobiology*, February 2009).

The author suggests that the human brain has capacities not found in monkeys. Their brains differ in meaningful ways, for example, in the proportions of different regions and the microstructure.

Hence, it is important to name his findings:

-human brain is 4.8 times the size of a hypothetical monkey of the same body weight;

-the gap between the human brain and that of the macaque monkey is twice as large as the gap between the monkey and a small insectivore such as shrew;

-neocortex is 35% larger;

-prefrontal cortex forms 28.5% neocortex in the human brain but only 11.3% in the macaque brain;

-the frontal polar cortex, area 10, is proportionally twice as large compared with a chimpanzee;

-consequential changes in microstructure that show that maximum spine density of layer III pyramidal neurons in the prefrontal cortex is 70% greater in the human brain than in the macaque brain;

-very significant differences occur between the column width in the left that is 17% larger than in the right hemispheres of the human brain, but not such asymmetry occurs in the macaque brain;

-are more magnopyramidal cells in the left rather than in the right superior temporal cortex, and in the left rather than in the right Broca's area, which is the change not noticed in monkeys;

- paracingulate area 32 of the human brain has no homolog in the macaque brain;

-spindle cells (or von Economo neurons) in the anterior cingulate cortex and anterior insula of the human brain do not exist in the macaque brain.

All mentioned distinctions developed quite recently (probably in the last 50,000 years or so) and are related to the rounding process of a human skull and the consequent generation of language. The rounding process reflects an organizational and functional change within the structures of the human brain. All such changes significantly contributed to produce the language-ready brain.

In this part of an extensive *Introduction,* I am trying to explain how humans solved "*dimensionality*" by reducing everything to linear workability. Such reduction also and significantly affected the generation of language.

A discussion on "natural bridges of vocalization"

Our brains can compute three space dimensions, extracting the elements used for comparing. It occurs a rapport established between various properties (like length, weight, etc.). Something is twice as long or twice as heavy as another. In this case, that "another" becomes the "base unit of measurement."

In language, various vocalizations turned compared according to their physical properties. In the practice of speech were used natural **codimensionality**, **covariation**, and **correlation**. These tools are "**bridges**," which naturally exist between the vocalized elements.

Stanford W. Gregory in *Analysis of fundamental frequency reveals covariance in interview partners' speech* published in Journal of Nonverbal Behavior 14, 237-251 (1990) indicates:

"*Spectral analysis of the fundamental frequency band in speech reveals covariance of voice energy levels and thus a possible form of rudimentary social synchrony. Though a number of researchers have observed this phenomenon of paralinguistic covariance, techniques for examining it have not been exploited...*"

"*This study reports on research showing that the **acoustic signal conveying covariance information** resides in the fundamental frequency band of the speech spectrum...*"

"*Research conclusions present an efficient, lucid, and reliable method for analyzing the paralinguistic mode of nonverbal behavior, and, in addition, offers evidence that the nonverbal, vocal channel of communication carries a signal embodying a semantic message*" (that regards the meaning).

Language studies, where the **language is a complex system with multiple dimensions**, demonstrate **emergent dimensionality** across development. *Out of the interactions between the individual elements in the systems, behavior emerges at the system's level as a whole. This so-called higher-order behavior cannot simply be derived by aggregating behavior at the elements' level.* (The previous paragraph was quoted from the paper *A review of common characteristics of complex systems* distributed by the University of Groningen).

Here vocabulary, grammar, narrative, discourse have their dimensions developed as independent abilities.

Lucio Biggiero of University La Sapienza (Rome), in May 1998 stated:

"*The **semiotic complexity** represents the infinite possible interpretations of signs and facts, while the **semantic complexity** consists of an infinite possible interpretation of words and texts. Artificial, natural, biological and human systems are characterized by **the influence of different sources of complexity**, turning them more complex*".

The apparent dimensionality of different language systems would mostly be emergent rather than innate.

Many linguists use a correlation matrix in their studies because linguistic relationships become more robust when the correlation coefficient is further away from zero. Correlation is a scaled covariation.

Covariance is employed as a statistical measure of the directional relationship between two elements or assets in finances. It is used to **reduce risk by providing diversification**. Hence, **diversification produces an uncertainty reduction**.

One may consider the language as a toolset. But such a toolset assembles things similarly with every known type of technology because it exploits the natural properties embedded in the matter's fabric. For this sake, *all technologies are a disguised copy of our language*.

Every language can generate an infinity of other languages. Technology can produce an infinity of various technologies by playing with its assembling/disassembling capacity.

Verbal and its utterance are not the only language: here, there are the sign, visual, musical, mathematical, and many more types of languages. All these languages use the same assembling and disassembling process, known more generally as analysis and synthesis.

Is language a technology?

As *Professor Dr. Jon Dron* with Athabasca University, Edmonton, Canada, has stated in a seminal paper with the title "*Is language a technology?*":

"*We can use language to manipulate ideas, create and transform concepts, design, explore, analyze, and more to achieve some goal or goals. We can use language to manipulate language, to construct things in language, and use those constructions to make other constructions*".

"*It is in many ways the fundamental human invention, hugely more important than everything else, like fire, the wheel, or the Internet. To* be human without language is barely conceivable; a lack of language to become widespread, we will no longer be humans".

"Language is the ability that enables us to symbolize, abstract, construct, amplify, and enhance, but also it generates the opportunity to spread technologies, build ideas, learn, create, discover."

"It's a wonderful virtuous circle that leads to an ever-expanding explosion of knowledge in our species as a whole even though we, as individuals, are likely getting dumber and are very likely dumber than some of our distant extinct cousins. It is not intelligence that makes so successful as a species: it is how we use technology to amplify that intelligence".

Assembling combines several simple elements, creating complexity, where the original properties turned significantly changed, while some properties become amplified.

The same mentioned "*bridges*" are those that naturally help assemble a complexity. But here, the process does not employ the "*reduction mechanism.*"

It scales up from the large macromolecular scales, where the number of coordinates is minimal, and behaviors are very simple, toward small scales, where the number of components is maximal. The entropy corresponds to the complexity at the smallest scale. The scale of the behavior equals the number of components involved in that behavior.

A large-scale complexity (macromolecular) requires "*order*" because it coordinates many small-scale components.

One can say that a large-scale complexity (macromolecular) tends toward linear behaviors; the small-scale complexity is fundamentally nonlinear, and the bifurcation points abound.

The high complexity of smaller scales leads to "*adaptable behaviors*" because its relationships must comfort a variety of many components.

The low complexity of large scales (often macromolecular) displays "*efficient behaviors*" because they are linearly maneuvered and tend to stabilize.

Complexity increases as the scale decrease. But every coherent system has the same amount of complexity on all scales.

We observe increases in most complexities surrounding us, and we do not realize that their increasing unpredictability is proportional to the rise in nonlinear behaviors.

Our brain cannot apply a dimensional reduction to those ever-growing complexities, and no one would know in what direction they may evolve. Many are scared by such growing complexities.

Language brought us two distinct natural phenomena: the scaling up of complexification that caused "*intelligence*," which is nonlinear, and simultaneously another phenomenon reduces the dimensionally of high complexities to low dimensional ones, making them be perceived as linear and conscient outputs.

Eventually, "*mental intelligence*" output is projected to the unconscious level. Constantly, the brain reduces such nonlinear products' dimensionality and brings many of them into the linear realm.

"*Inner intelligence,*" being produced in our brain, manifests as intermittent pulses of variable intensity. The intensity depends on the number of elements implied by each idea. For example, a multidisciplinary approach would benefit from a larger seeding and deliver an emergent result (concept) superior to each of the particular field's ideas used.

"*Outer intelligence*" of products exposes a stable and continual accumulation of exterior elements. Contrasting mental outcomes, such continuity of "*outer intelligence*" assemble more complex products with a significantly larger number of components. The "inner intelligence" is far distant from matching such considerable larger outer complexities.

Humans before language were very similar to all other animals. *Such different animals could climb a few steps up on the scale of intelligent evolution.* But *language made us rise to the top and continues to amplify such a distinction.*

One may ask if mental complexification could drive the unconscious thought into higher hierarchies? The unconscious thought, being nonlinear, remains undisclosed and unknown to the conscient thought.

Even then, it could be influential to some extent because it is projected back into our consciousness as a dimensional reduction. Indeed, it would enlarge the conscious output in a linear form. I would say that *"intelligence" is in the language that is transmitted to technology*. In other words, **"intelligence" is the result of "assembling/disassembling."** *Intelligence naturally resides in "complexities" that result from "assembling/reassembling."*

Emergent behavior may be also intimately correlated to those aforementioned "bridges". Eventually, **the process of combining into complexities must relays on "codimensionality" and "covariance", which features will avoid the noise that can compromise and prevent the emergent to occur**. Hence, in this eventuality, an emergent behavior is a scaling up that characterizes a higher dimensional space in a higher hierarchy that must be nonlinear, hard to define or understand, and unpredictable.

Implicitly, in such a case, the "intelligence" is nonlinear and equivalent to an "unconscious thought." It is physically invisible, being determined only by its behavior that defines a system's reaction to a probability.

A complex system generated by initial flexibility needs a certain level of variety caused by **differentiation** because it **introduces stability**. Even when the dimensionality is reduced by differentiation, causing a loss of information, by collecting those differentiated parts, novel self-organization intervenes to produce an emergent property of the system, a property that the individual components do not have.

The **"intelligence" is an emergent property of that complexifying system and results from the self-organization of components; thus, it is not a mental feature developed by the brain**.

Hence, the **differentiated parts' assembling produces a** *collective "intelligence"* in the form of a novel but an emergent property of the whole. Novel high-level properties make that system behave intelligently. They are directly generated by collective dynamics of the nonlinear interactions among the components. The new emergent order appears like a macro result of the nonlinear micro-interactions.

Every complexity tends to grow exponentially, increasing a type of chaotic evolution that opposes "order." The complexification in dynamic-adaptable systems needs additional energy to minimize chaotic growth and preserve a form of equilibrium. All biological systems are negentropic. Negative feedback loops maintain the system equilibrium. If the energy needs are not appropriately fulfilled due to intervening positive feedback loops, the system falls apart.

As it appears, the **"intelligence," derived from complexities, naturally tends to increase exponentially**. But this does not characterize a mental state or process in itself: **it indicates that the mentioned process continues to evolve exponentially on its own, producing an exponential augmentation of newly generated complexities**, which become the subjects provided to further disassembling/assembling operations. Hence, it is nothing mental about such evolution. It is there a natural mechanism that works until it breaks apart, crossing into chaos.

In this case, the brain is not stimulated to evolve because the **mentioned mechanism independently acts on its terms**. By contrast, the brain deals with a continual reduction in initial information that reduces the nonlinear quality of processed data. It diminishes the initial meaning and causes progressive dumbness.

I like to quote James Ladyman, James Lambert, and Karoline Wiessner with a significant example in their paper *What is a Complex System*? They explain:

"For example, when they undertake complex tasks, the ants behave as they do because of the way they interact with each other. On the other hand, no individual ant has any idea what they are doing, and left to their own, they will exhibit much simple behavior."

As some scientists see it, we have turned dumber in the last century, while our power of assembling increased; even then, we are capable to "assemble" an ever-increasing number of new technologies. As the mentioned process evolves exponentially, in each step, we lose some of the initial information. This makes our brain lose access to prime data, and thus, the processing becomes increasingly dumber.

Method and materials

I used in this paper the data provided by the *Information Theory* and many associated ideas. I corroborated the mentioned data with neuronal biology and information on the neurogenesis process.

I also used Complex System Theory and associated ideas on complex behaviors.

This paper was initially motivated by *Professor Jon Drone*'s publications from Athabasca University, Edmonton, Canada (as mentioned in references).

I used information, as quoted, from *Professor Salikoko S. Mufwene* of the University of Chicago. His article I mentioned was included in the paper *In Search of Universal Grammar: From Norse to Zoque*, edited by Terje Lohndal, John Benjamins.

I have also quoted *Dr. Srishti Saha* on the connection that exists between covariance and correlation because it shows how the mechanism of language functions at its fundament.

I quoted James Ladyman, James Lambert, and Karoline Wiessner from the University of Bristol, UK, with the paper *What is a Complex System?*

I used *Motor prediction,* written by Daniel M. Wolpert and J. Randall Flanagan, and published in Primer paper.

The science of complexity research shows that each trillion of highly specialized cells (neurons) starts as a single primordial cell. How undifferentiated cells commit to their biological fate, causing differentiation?

The **research** indicates that cells face multiple competing choices and perform binary decisions between competing groups of genes. In this competing genetic process, it appears a genetic covariance that is responsible for a gradual biasing toward a more robust genetic program that affects the cell ultimate commitment, as it produces a specialization.

By a linear **dimensionality reduction algorithm**, as nature produces its version, the entire datasets are projected on a linear space, where the dimensions are prioritized to explain the variance data.

Covariance measures how much two random variables vary together within a populace of variables.

In our Mathematics, a matrix describes the relationship between different dimensions existing within a system. One can assume that there are various sorts of matrix analysis in nature, analyzing the relationship between system's dimensions. Each matrix has hidden factors, which uncover universal properties. Each matrix can be decomposed, unveiling the fundamental properties of those dimensions.

Here again, the **choices are determined by natural codimensionality and covariance, which influence the genetic competition**.

However, I have to mention that most investigations are based on variance and covariance equation analysis in chemistry. For example, the intensity of segregation (degree of mixing) is defined as the covariance fluctuations between two chemical species.

Haiping Huang, in his study *Mechanisms of dimensionality reduction and decorrelation in deep neural networks*, published by arXiv on November 27, 2018, indicates that the author "*proposes **a mean-field theory of input dimensionality reduction in deep neural networks***" that also shows how covariance level (redundancy) varies along the hierarchy.

A deep network is a multi-layered neural network performing hierarchical nonlinear transformation of sensory inputs—the **dimensionality reduction results from nested nonlinear transformations of input data**. Covariance is considered a tool for determining the constitutive linear structure that results from the reduction process.

The results

The results indicate that neurogenesis can explain the facts that made the language behave like technology. I associated to neuroscience data the Complex Systems and Information Theory information.

Let's see how Mathematics transforms a nonlinear product into a linear result.

Here, there are two steps:

-in the first step, the nonlinear input data is transformed into a higher dimensional space by using a **nonlinear mapping**;

-the second step generates a linear separating **hyperplane**; a hyperplane is a subspace whose dimension is one dimension less than that of its ambient space.

As Wikipedia explains, "*the difference between a subspace S and its ambient space X is known as the **codimension** of S concerning X. Therefore, a necessary condition for S to be a hyperplane in X is for S to have codimension one in X.*"

I would say that a language-ready brain became possible when anatomical changes and transformations **solely allowed the cognition, but not the olfaction, visual or other senses, to identify codimensionality, covariances, and correlations**. Only such features, being used as natural bridges, permitted language development by operating mandatory disassembling/assembling, analysis/synthesis.

In neural networks, the neuroscientists see a similar transformation path, where a **neural type of hyperplane** generates a nonlinear mental mapping that may qualify for unconscious nonlinear structures.

Such a process is said to occur in all types of brains.

Let's see the three stages of human neurogenesis that bring nonlinearity down to the linearity of consciousness.

The **first stage** of neurogenesis produces **undifferentiated neurons**, which can adjust to almost every challenge. These unidentate neurons represent a **sum of all probabilities** that exist. They behave nonlinearly and have **no structure**.

In the **second stage**, the brain prompts to make a selection out of such undifferentiation. In this second stage, it is generated the **mental mapping** that contains **nonlinear structures**.

In the **third stage**, the <u>**cognitive mapping**</u> **of nonlinear structures is transformed into linear structures**. This last step generates consciously expressed linear outcomes.

Overall, the mechanism switches from nonlinear quality into linear quantity by lowering the dimensionality. The process mandates the existence of a codimension that intermediates the translation.

For example, the first defined quality probably was that of daylight. Several days are complete by a multiplication process, designing a cycle named week, month, or year.

The concept of a day was nonlinear and multidimensional, too. By reducing this multidimensional detail, a detailed quantification was created like that of the hour. The dimensionality of an hour was further reduced, making minutes or seconds.

Each of these subdivisions is produced by the introduction of a mental **hyperplane processing**. Each subdivision is one dimension less than the original product. Each subdivision represents a disassembling that reduces the actual variance that is the original information.

Combining the disassembled pieces, the process does not build the original nonlinear dimensionality of the whole again. The <u>*reconstruction error*</u> *(information loss) is defined as the mean squared distance between the original and the reconstructed points*.

Dimensionality reduction comes at the cost of information loss.

That is why the sum of the parts is never equal to the nonlinear whole.

But it matches the whole when we switch to linearity. As it seems, our brain and mathematics cannot restore an initial nonlinear dimensionality that existed before hyperplane processing occurred.

However, mental selection processes reach those elements of reality, which possess minimal variability, indicating a slow rate of change, which can be mentally computed, controlled, and predicted.

All **low change phenomena and processes are linear and possess a maximal covariance level**, allowing "bridging" and manipulation.

A linear space enables the brain to <u>compare the elements</u> using the "bridges" because <u>covariance is maximal when the bond is linear</u>.

Linear selection suffices to increase the rate of survivability because it **provides prediction, control, causality**, which all result from "bridging." In all, such a linear selection is a short-term and fast approach required by the immediateness.

Discussion of results

I could assume that the undifferentiated neuronal stage of neurogenesis would be familiar to all animals. It provides a potentiality for all probabilities. It is fully nonlinear, but no one on this planet can consciously compute nonlinearity.

As is the case, our brain and all animal brains could process nonlinear data, but the result remains unconscious. Even then, it is known that the unconscious processing provides much of the data expressed consciously. This type of transfer is caused by those **bridges** mentioned above that **allow selection/differentiation**.

In a nonlinear data structure, such data is hierarchically connected and is present at various levels. These nonlinear structures use memory very efficiently. The data remains the same, disregarding an increase in its amount.

However, linear data structures are not memory friendly and cannot use memory efficiently. When this data's complexity increases, the size of the number of data increases, too, causing a significant burden for memory accumulation.

The intuition is nonlinear, introducing a correlative approach. Our ancestors may have confused the **correlations** they made for a natural type of causality. But here, *Dr. Ciaran Lee*, Senior Researcher Scientist at Babylon and Honorary Senior Research Associate at UCL, stated the following: "*We combined multiple correlating variables from* incomplete medical datasets and showed, with a high degree of confidence, which correlations mean causation."

Several other researchers in Artificial Intelligence found *causative variables* and identified *correlative causation*. Hence, if our ancestors were somehow wrong about the correlations, that thing was correct only in part of their estimates. Such a partial correlativity occurs because it exploits only one or more codimensionality. The rest does not show *codimensionality*, and therefore, it does not fulfill a causal correlation.

In Mathematics, nonlinear correlations are considered fallacious. But in Genetics, some new methods are used to quantify a global nonlinear relationship; and, they find evidence for *some local nonlinear correlations*. Probably again, such a case arises because of a limited codimensionality.

A minimal linear selection should occur in all animal brains because only such a selection can deal with the surviving needs. The surviving choice stands for the reduction mechanism that introduced linearity into the animal world.

So, the linear section always existed because it defines surviving and is always intermediated by the aforementioned **natural bridges** embedded in the fabric of matter.

Hence, all animals would have a minimum of consciousness, and it would differ from one species to another.

All animals process the bulk of perceived data by the olfactory system and other senses, which are nonlinear. Their limited developed cognition (because their neurogenesis output does not support it) selects only a small amount of structured data that refers to minimal communication, minimal control, and planning. This minimal selection is bridged to their conscious processing, becoming structured.

Because of their switched type of neurogenesis, cognition processing extensively prevails in humans, while it provides an enormous increase of structured data. All such structured data turn conscious. The same structured data becomes available for communication, planning, control, and socialization, which functions are hugely developed in humans compared with the rest of the animals.

There is a clear distinction between very many selections produced by cognition and a significantly much smaller amount provided by olfaction and other senses. These senses have little room for development.

The neural correlations of a cognitive map are speculated to exist in the hippocampus and entorhinal cortex cells. These correlates make cognitive maps appear significantly more extensive in humans compared with the rest of animals.

The **cognition behaves with extreme plasticity** because it remains open to vast unstructured incoming information while gradually structuring it. **The plastic function is nonlinear**, and its nonlinear rules are addressed by some **neural receptive fields** displayed hierarchically.

By structuring, the mental outcome moves down the hierarchy by several consecutive dimensional reductions, while ultimately, it encounters a **linear adaptation**.

From language practice, everything else turned dissembled and reassembled, making products for daily needs, which, at the moment, have increased the species' immediate survivability.

As it appears, humans survived extinction by acquiring the language ability. This ability came in parallel with a biological adaption to various challenges, which together prompted African migration.

But here, the **Theory of Complex Adaptive Systems** tells a different story.

Efficiency would have a complexity profile where it resides a **lower complexity** that extends on large scales. In creating a large-scale complexity, the **lower-scale complexity** was sacrificed and **resulted in non-adaptive trends**. The small-scale, based on small communities of around 150 individuals, shows efficiency, is sustainable, and more functional at the ecological level because knowledge is based on personal experience. They had knowledge based on relationships. They made interaction among species and environmental factors more personal.

In Africa, such a situation of non-adaptability is well reflected by the case that only a single Y-haplogroup (haplogroup A) was generated for at least 100,000 years. The second Y-haplogroup (haplogroup B) appeared just before the out-of-Africa migration or around 70,000 years ago.

These Africans, mostly known to live in many ocean-shore caves located at South Africa's peak, migrated to a tropical East Africa only cca. 70,000-65,000 years ago, generating mitochondrial haplogroup L1, L2, and L3. These haplogroups were those that later spread around the world.

Before they migrated to East Africa, these people were not adapted to challenges posed by out-of-Africa pathogens and other various environmental constraints. They were not adaptable to unforeseen variations occurring within their biology and within the environment. Hence, their repeated attempts to leave the continent eventually produced devastating results. It is known that several waves of African migration dead-ended in the Levant and others in South India and South China. It is to assume that many of the early migrations, at some point, reached extinction.

In Africa, neural complexity evolved into an "*efficient*" type that enhanced long-term neuronal stability. This stalemate situation was broken about **70,000 years ago** when a faint manifestation of art reflects an intervening novel neural specialization. A burst in these people's neurogenesis can be connected to a massive geophysical event (geomagnetic excursion) that lasted from 80,000 to 70,000 years ago (the Norwegian-Greenland Sea Geomagnetic Excursion).

Such a situation indicates that a major switching in human neural activity occurred just before archaeologists and geneticists dated an African migration out of the continent that initially entrained no more than 1,500 founding males and females.

Consequently, one **can suggest significant switching in human neurogenesis that manifested at full scale only 10,000 years later, around 60-50,000 years ago.**

The assumed switching indicates that neurogenesis increased suddenly, causing a **rapid augmentation in the production of undifferentiated neurons**, which show greater complexity and produced higher neural plasticity. The new mental setup sacrificed its original large-scale behaviors, **reducing individual complexity**. Immigrants started to do things much easier, being released from a local group or tribal pressure and constraints.

On the other hand, the switching allowed (large-scale) open-minded cooperation between communities. The **neural efficiency was replaced by high neural adaptability to changing conditions and distinct possibilities**.

As the Africans moved out of their continent, the challenges caused by new, very distinct environments incited an additional mental stimulus that forced them to developmental complexities of the size, at least equal, to the problems to be solved.

It was documented that groups being settled in South and Southeast Asia migrated back into Africa, which in fact homogenized the African genetic material with the rest of the out-of-Africa migrants.

However, the Complex Adaptive Systems Theory stipulates that the said system may fluctuate over time, switching between the mentioned two states. Our history is full of such behavioral fluctuations, changing from grand cultures, expressing efficiency and stability, to their total alienation caused by conquering migratory people, who came with higher brain plasticity and adaptability.

The complexity states may cyclically oscillate between "efficiency" (stability) and "adaptability" (flexibility).

It has been argued that "*language is a natural collective technology that evolved primarily to facilitate efficient communication in populations*" (Prof. Salikoko S. Mufwene, University of Chicago). It was found that language implies several other things, which are as important as the communication itself: they are **planning** (all utterance is designed in mind), **control** (the reason of language is to control the audience, and by such authority, the statement can be shared), **socialization** (there is no reason one to talk for himself only).

In my opinion, the **essential clue implied by language acquisition was a new capacity to structure data through the only available venue that was provided by cognition**. The other senses allow limited plasticity, and they could not support the size of the complexity, like that imposed on cognition by the advent of language.

The mental complexification that occurred (due to language, migration, and complexifying) helped with **inner and outer intelligence**, where the last became the maker of technologies.

The leap forward started some 60-50,000 years ago, probably extending the "mental mapping," but the language revolution occurred only 40-30,000 years ago when a chain of other geophysical events favored another increase in human neurogenesis followed by a massive linear conversion.

I have to note that most specialists estimate, based on limited experiments on feral individuals, that a languageless person would have a very primitive consciousness, with weak memory and limited cognition.

However, this limited study shows the **eventuality of losing language (at an early age) would inflict significant functional changes in the brain**.

Modern people think the way they talk. Not talking and not hearing any talking around you will influence your thinking process to adapt differently to such a non-communication condition. These individuals made for surviving a mentally driven selection based on non-communication approaches, even when their neural network was genetically set up otherwise.

Human language development may downplay other senses output by diminishing a particular type of traffic on multiply used circuits. **Language has displaced or interrupted other forms of thoughts and interaction**, like those based on olfaction and emotion.

Language development significantly enlarged socialization, prompting self-domestication that altered the old emotional aspect that had primarily driven a prehistoric person's reasoning. It improved human capacity to adapt to new environments and conditions, prompting a migration around the planet successfully. Neural specialization contributed to significant memory, learning, and cognition improvements, gradually building up the modern brain.

Conclusion

However, the disassembling/assembling process in our brain led to a partial loss of initial information.

A similar process of information losses was translated into the buildup of technologies.

Initially, the reduction diminished the environmental uncertainty associated with cognition (beliefs and attitudes) and behaviors. It switched to linear uncertainty that hides behind an information loss.

The technology worked as an extension of the language. It introduced the predictability and direct control over the uncertain inhabiting the natural outcomes. It introduced processes and functions with predictable/controllable results.

A car shows an entire function different from each of its assembled elements. Hence, an artificial complexity has an overall quality that is higher than each of its components. It is the "*inner intelligence*" inhabiting a particular "*idea*" that becomes projected into the "*exterior intelligence*" of the said product.

However, *the product of artificial complexity is more intelligent than the idea that created it*.

Nevertheless, all products have significant information losses from initial data. As technology developed, the information losses proportionally augmented, instantiating us further away from natural reality.

We build upon our imaginary products, where each product behaves more intelligently than its creator. One would say that, in nature, the characteristics of complexity are driven by an adaptation to continuing changes in the ambient. The components of that complexity develop nonlinear relationships, where they behave as self-organizing; the relationships tend to generate an optimal design that fits the ambient.

In our human world, *our products cannot self-evolve*, yet

AI would be "intelligent," too, and again, more intelligent than us, but we would need to create self-evolving features.

* * *

As it seems today, the initial **individual uncertainties** were communicated within groups or collectivities at an ever-extended level, **becoming social uncertainties**. Hence, personal uncertainty was reduced only at an apparent and temporal level, serving the moment's surviving needs.

Technology introduced a contaminant concentration ratio that exponentially accelerates as more technology is created and displayed within the environment. These contaminants are linear and profoundly change the nonlinear habitats and ecosystems, devastating their original qualities and compromising their nonlinear life-supporting capabilities.

Such linear contaminants vastly affect the surroundings because they interact unobservably with one another in ways far from our understanding.

We have these linear contaminants because we cannot restore the initial dimensionality without errors, obstructing natural recycling.

As many scientists admit today, our technology-end-result **increased the uncertainty in an existentially hazardous manner**.

Acknowledgment

This paper aims to explain how language was produced by enhanced neurogenesis and why it embeds all technology features.

References

Salikoko S. Mufwene-*Language as technology: some questions that evolutionary linguists should address*.

Jon Dron-*Is language a technology?* Published in Athabasca University Landing, September 5, 2010

Stanford W. Gregory *Analysis of fundamental frequency reveals covariation in interview partners' speech*, published December 1990 in Journal of Nonverbal Behavior **14**, 237-251 (1990)

Seven Papers

Gleb Kumichev published the paper *Dimensionality reduction with PCA: from basic to full derivation* in Toward Data Science on April 13, 2020

Hindawi/2020: *An Introduction to Complex Systems Science and Applications*

Daniel M. Wolpert and J. Randall Flanagan: *Motor prediction*, in Primer

James Ladyman, James Lambert (Dept. of Philosophy, University of Bristol, UK), and Karoline Wiessner (Dept. of Mathematics, University of Bristol), published March 8, 2012: *What is a Complex System?*

Hindawi/2020: *An Introduction to Complex System Science and its Applications*

A review of common characteristics of complex systems as a paper distributed by the University of Groningen

Lucio Biggiero with *Sources of Complexity in Human Systems* published by University La Sapienza (Rome), 1998

Haiping Huang with the study *Mechanisms of dimensionality reduction and decorrelation in deep neural networks*, published by arXiv:1710.01467v3 (cs. LG) November 27, 2018

Gerald Wolff with *Information Compression as a Unifying Principle in Human Learning, Perception, and Cognition*, published in Volume 2019, Article ID 187946 on February 20, 2019

Richard Passingham with *How good is the macaque monkey model of the human brain?* published in Current opinion in neurobiology 2009, Feb; 19(1); 6-11

Chris Mitchell and Scott Nash, with *The Intermixed-Blocked Effect in Human Perceptual Learning, Is Not the Consequence of Trial Space* published by Journal of Experimental Psychology Learning, Memory, Cognition, 2008, Vol. 34, no 1, 237342.

Dan M. Mrejeru on academia.edu *Hypothesis of Homo loquens* (2020)

Dan M. Mrejeru, academia.edu *Absorption of C14 isotopes stimulates neurogenesis and biophoton production* (2020)

Dan M. Mrejeru, on academia.edu, *Changes in brain anatomy inflicted the development of language.*

Language brought us the intelligence of complexification

The language is based on continual processes of assembling and disassembling.

In the meantime, the Science of Complex Systems shows that complexity results from assembling. In nature, the assembling effects the components' free-interaction; it may be caused by the relationship developed by interacting components or parts. In our human world, we deal with our artificial type of assembling planned and thus human-made. But both ways cause evolving complexities.

It is paramount to state that when the language developed in the human brain, it attuned us to one of the most proficient features encoded in the universe's natural fabric: ***complexification***.

It allowed us to produce our own type of complexification. No one in the animal world, except us, did ever have contact while progressively using such a central natural feature.

The Science of Complexity also shows that the assemblage, causing complexity, generates a novel feature that does not exist in any of the assembled parts. This emergent feature is the ***intelligence*** of that system.

Now, it becomes clear that the ***language-ready-brain*** allowed us to produce complexifications, which contain intelligent features.

When one got the language, one began to think about the way he/she talked. This new way of processing the thought changed the thinking process itself, making one assemble his/her thoughts how one gathered the words in the language.

We entered a new stage in evolution that is about complexifying with an emergent/intelligent behavior added to it.

Here, it may be the case that a geophysical event placed 80,000-70,000 years ago initiated a change in Homo sapiens brain by favoring a higher than average neurogenesis. It produced a larger than the regular number of undifferentiated cells. The process, at a sudden, generated an opportunity for the undifferentiated cells to adapt by the process of specialization/differentiation.

Somehow, the mentioned event allowed Homo sapiens of East Africa to achieve a mental complexity that matched the complexity of environments placed outside Africa. In other words, it drove the Homo sapiens to a level of effectiveness that matched the scale of those environments.

Before this time, the Africans made many tentative to exit their African environment, but they failed because they were not mentally compelling enough.

Thus, Homo sapiens were able to leave Africa but were not having a language-ready brain. What led to such a novel brain?

The science of complexity may suggest that the migrants, being challenged by new environments with new conditions, were subject to inevitable stress that produced gradual new adaptation, **forcing their brain to expand the response capacity to the size of the encountered problems**. The migration process led to a mental complexifying. It was the beginning of the process of complexifying.

This mental expansion gradually created novel but larger mental complexities. The **complexifying process inflicted changes in the brain anatomy and circuitry**, reflected in the skull shape's rounding. When this last detail was accomplished, **around 30,000 years ago, the language-ready brain was set to initiate a language revolution**.

The language on its own initiated the production of **linguistic complexities as generators of intelligence**. Finally, such novel intelligence assured the migration progress, allowing the migrants to successfully settle in various regions.

Hence, the **Homo sapiens turned equipped with complexifying behaviors, which made him produce an exterior intelligence added on top of the cognitive intelligence embedded in the language's practice**. It represented a fundamental change that never occurred in any of the other animals.

Language caused a fast evolution. Henry C. Harpending of the University of Utah and John Hawks of the University of Wisconsin estimated in their study published in 2007 that "*over the past 10,000 years humans have evolved as much as 100 times faster than at any other time since the split of the earliest hominid ancestors of modern chimpanzees.*" They found that "*at least 7 percent of human genes underwent evolution as recent as 5,000 years ago.*" Much of the change involved adaptations to particular environments and conditions.

Such changes are reflected by the enlargement of several cortical features, like the parietal lobe and the cerebellum's bulging. It also implies changes in many brain components.

Several recent studies using fMRI and other novel medical technologies found that cognition uses in language processing specific pattern-identification products generated by various brain regions, which are not known to be related to language.

This indication unveils the double uses of various other sensory neural sections and pathways for language acquisition and expression. It **implies a significant increase in the internal connectedness of the brain**. A similar result is provided by other studies, where the researchers found visual circuits being intensively used by language processing.

All the above aspects are complexification-related. The research also discloses some novel combinatory operations responsible for creating the sentence as a hierarchical structure of words. Further on, it implies that human brain lateralization is most consistent, directed, and pronounced than other mammals. This highly evolved asymmetry is due to language development, too.

The complexification caused by language may affect the plasticity manifested in the interrelationships between various neural structures and their components.

Maybe we live with a confusing idea about **what intelligence is?** But scientifically, **it is only one thing: the emergent aspect arising from complexification**. And hence, it consistently defines intelligence as a continually produced but emergent feature. It is not innate.

The most common way to define human intelligence is the IQ test. Using the idea developed above, the high/low-scores of IQ may result in each individual's assembling/disassembling capacity. The emergent result of the mental assembling/disassembling helps produce a complexity equal or not to the complexity of the problem; according to such result, it may help or not that problem-solving.

Let me detail this aspect of assembling that emerges in intelligent behavior.

When one makes a sentence, one assembles words. The sentence acquires an emergent meaning that is not present in any of the terms used in assembling. The sentence results in an emergent/intelligent feature that was not proper or innate to the assembler. Again, it arises solely from the complexification process.

Each sentence, as an assemblage, brings a bit of natural intelligence that does not exist before this process occurred. This first step made the language users cause the assembler's inner intelligence and the exterior intelligence of complexifying. Both types of intelligence solely resulted from language.

Further on, phrases and all other linguistic complexities developed by the language automatically caused more intelligence in the outcome.

At the societal level, when one introduces in mental analysis other components, like ethics, multi-racial, multi-cultural approaches, and so on, the complexification process increases by the additions made and the result changes within each addition. The result becomes more intelligent.

However, the populations at large are not educated to confront such additions, tending toward a reductionist model, where the complexity is low. It is now termed a "populist" movement.

In general, reductionism causes a disassembling and elimination of certain elements, lowering the assembling complexity while favoring a low emergent/intelligent outcome.

However, the complexification can extend beyond any known borders, while at some point, it would dismantle into chaos. Such a collapse will reduce the complexity back to much fewer elements and interactions.

The collapse may have two origins: one is related to a limiting supply of energy, the other one depends on internal connectedness. When the inner becomes more flexible by increases in interconnectedness, it races the system to acquire more overall energy, crossing the energy limit, leading to a collapse. Our actions make systems more susceptible to collapse.

The complex systems appear to have a life of their own because their inner relationship responds to changes in the environment. Such a response is not according to our intentions and expectations, and it fails our predictability and understanding; it denies any certainty.

As Joseph Tainter in *The Collapse of Complex Societies* explained, the increase in complexity comes from our attempt to solve problems. It turns out that each solution we apply introduces unintended consequences, generating new problems. Every benefit is associated with its cost. One gains something by losing something else. Sometimes the price is larger than the gain.

More generally, each time we solve a local problem, we create a global one worse than a locally solved one. And the more complex, the greater would be the fail when things go wrong.

Hence, we are racing to increase the complexity of problems and the complexity of solutions. The benefits gained by solving problems by introducing more complexity, in the end, cause diminishing returns and negative results.

Today, many specialists see complexification as a growing danger. Somehow, this idea becomes closer to reality at the technological level, where development is exponential, and the emergence/intelligence inured in each new product tends to generate a general result that is way over our level of understanding.

In fact, one deals here with a **_technological intelligence gap_** that enormously increases between the producers and their products. And it could be hazardous.

People do not realize that **_our intelligence is not innate, but it is continually and emergently produced_**. Our mental state remains low because the mind does not collect or accumulate such emergent/intelligent results except by a limited human memory. But, we collect all results by recording them exterior to the mind. It is an external feature. But each mental pulse remains a one time-deal with no genetic influence and little cognitive improvement.

Regarding **Artificial Intelligence**, I would say that it benefits from **_accumulating positive results from one step to another_**. This accumulation is even significantly higher than anything manifested in our other technologies. **AI technological intelligence gap gradually becomes higher by day**.

All complexities naturally evolve toward an entropic collapse. Probably, the fee-interaction between its components or parts reaches a level beyond natural computation can deal with. It is a point of no natural resolution.

Another aspect that concerns the self-organization process is a bottom-up process but where eventually the fundamental quantum level intervenes with a top-down regulatory approach. Our still infancy in understanding quantum mechanics prevents us from understanding such an ultimate quantum regulation.

Maybe quantum computing practice would bring more light to such fundamental features, like when, why, and what makes the assembling reach the disassembling point.

At present, most scientists consider that the limits of complexity are in its capacity for absorbing energy. Hence, according to this, the energy is the limit and is set up by the inner relationship's needs.

At the mental level, the brain's energetic memory needs are resolved by erasing some of the older memory in favor of new ones.

However, we solved this mental problem overall by adding exterior memory storage.

About 10,000 years ago, our brain encountered an energy crisis generated by the development of language. According to Stuttgart University scientists, the brain forcibly resolved it when it changed the learning's neuronal rate.

The brain responded with a circuitry signal rerouting accompanied by a change in neuronal firing frequency. The new processing diminished the firing speed rate, making the learning occur slower, and thus, it had cut on some of the energetic needs.

In the meantime, this process introduced a new limitation that referred to an increase in the amount of time needed to accomplish learning of any particular task. It slowed down our awareness and its speed-reaction to events and actions.

We became slow-learners that have societal and individual safety implications. It was the price paid for adopting complexification.

Would there be more energetic cuts to pay for our rampant developing complexification?

The loss of language causes the civilization to collapse

I would start this essay by quoting Earth Institute professor Ruth DeFries. She gave an interview (December 1, 2020) to State of the Planet about her book. *What would nature do? A guide for Our Uncertain Times.*

She said: "*Societies have organized their economies based largely on efficiency.*" It means that, along with evolution, we learned to produce the most with a minimum of resources.

This aspect is well-explained by the science of complexity that shows that efficiency kills diversity at its fundamental level. It kills adaptability.

In her interview, Professor DeFries explains how our civilization tended to find a single optimal solution for each problem encountered. By contrast, the **biota developed a diversity of solutions**, which can resolve an issue in different ways, assuring that species survivability. It shows that "*diversification implies investment in energy and materials, it keeps the options open, it provides a portfolio of possibilities for life to persist.*"

"*Diversity in human societies takes many forms. Not only the diversity of plants and animals that feed humanity, but the diversity of languages, cultures, and knowledge. The tendency of the modern world is to squash diversity in the interest of efficiency. But that diversity is our reservoir of options for an unpredictable future. Building redundant and diverse parts in designs is a matter of life and death.*"

"*Civilizations are complex systems. Complex systems go through cycles of growth, stagnation, breakdown or collapse, and renewal., while these cycles are repeating over in the long course.*"

The science of complexity says that the products that resulted from diversifying are assembled to generate more significant complexities. Hence, diversification is part of a more extensive process of complexifying.

Some linguists (like Richard Armstrong) "*warn that the loss of linguistic diversity is akin to the loss of biodiversity.*"

More than that, the loss of linguistic diversity equals the failure of the embedded cultures, traditions, and wisdom because each culture represented the local type of answer to environmental uncertainty. It created local biodiversity where everything was adapted to that locally given answer. Killing the biodiversity, we directly kill the intelligence inured in that diversity. Less diverse is less intelligent.

However, the <u>process of complexifying is the provider of "intelligence" for that system</u>. **A minorly diversified system is less intelligent than a more developed one, meaning that the first one's solutions are inferior to the second one, ultimately affecting each system's survivability**.

Within the practice of language, we learned to assemble, producing complexities. Here, we learned to mimic natural assembling while creating our artificial complexities. The motor of such human assembling originated in language. A recent paper published in *Science Advances* (December 2019) indicates that "*anatomical ingredients (as the laryngeal mechanism) for speech were present in our ancestors much earlier than 200,000 years ago, and probably for several millions of years.*"

The above information confirms that humans, like many other animals, for a very long time emit vocal sounds and modulate them in a manner that inured a meaning. Still, they did not have a mental complexity capable of assembling those vocalizations. It shows that an anatomical change in the brain neural network and circuitry evolved only in the last 40,000 years and created the language-ready brain.

Such a novel brain was ready-for-assembling the complexity of language, giving us, in parallel and simultaneously, the benefit of that **intelligence** generated by the mentioned complexity.

At present, our civilization tends to replace verbal communication with digital and visual data processing.

As the old wisdom is saying, an image bears in itself a thousand words. Why is that? Because we started with images only, and when the language intervened, we disassembled that image meaning into many words and associated definitions. It implied a verbal translation of the visual.

The eye-to-brain pathway is called the **opticoders** (the visual brain) connected to the back's occipital lobe. In contrast, the ear-to-brain pathway is called the **lexicoders** connected to Broca and Wernicke processing areas located in front and central positions.

The people with evolved opticoders system heavily depend on lexicoding. It is known that visual images are great for receiving knowledge, but they are not adequate to process and communicate knowledge. Our knowledge is entirely dependent on mental word-processing. This is so because the verbal assembling and disassembling process create knowledge. Such assembling generates the intelligence proper to complexity or complex systems.

Before language, the opticoders could not cause any mental advances because they cannot produce assembling and its associated intelligence. The immense advantage introduced by language was its capacity to assemble complexities. The loss of language will annihilate such an advantage that solely created our civilization.

The experiments on animals indicate that opticoders and the coders for other senses do not have a natural mechanism for assembling data.

In humans, **assembling becomes possible because such data is processed by cognition**. In contrast, in all other animals, the same information is processed by olfaction and other neural networks serving the senses.

Hence, **a return to opticoders, as the central processors of neural data, will remove our ability to assemble complexities. Then, we will lose our intelligence resulting from an assemblage of complexities**.

We stepped back to visualization with the written language but intimately associated with vocal pattern recognition. It was part of language processing.

Today, many think that our language is outdated and needs fundamental improvements. Such an idea is driven by the enormous extent and success of our technologies. But people do not realize that all technologies without distinction are created by an assembling that adds the intelligence of complexities to these technologies. Indirectly and directly, such mountings or assembling are driven by our verbal capabilities summarized by language.

Here, it is a very fine line that we do not have to cross.

Our new technologies tend to reverse the process by translating back from verbal into images, causing visual pattern recognition. But images are not a language; they can tell a story that we need to decode. In doing so, the visual patterns decoding does not use any language.

The images are mentally analyzed (cracked) nonlinearly and consciously interpreted into words. They are holistically processed and reanalyzed within a pattern recognition process but not translated into a verbal mode. It only reduces nonlinearity into linearity using the features of codimensionality. The image processing stays independent of language processing because it uses opticoders, not lexicoders.

The translation of an image into words occurs only when it is mandated. In general, it is hard to translate an image precisely into words. For a good part of that imagery, one cannot find the appropriate describing words because there are not enough codimension bridges.

Images are processed in the brain, primarily nonlinear, and in the same fashion, they were processed before the advent of language. The image processing appears to be at the roots of the unconscious while nonlinear thought. Only recently, the investigation of unconscious thought has been performed and identified accordingly. Its interpretation is novel. Would visual images stimulate unconscious thinking, reducing the conscious output?

Digitalization translates qualitative verbal meanings into their digital quantitative counterparts. Quality is primarily nonlinear, while quantity is fundamentally linear. Digitalization trends to translate much of our unconscious thought into a conscious output. But here, visualization tends to do the opposite. Would there be an association between these opposing hypostases?

In her article (November 14, 2018) on Diverse, Tanya Leake shows that *"communication is only 7 percent of what we say verbally. Even conservative research measures nonverbal communication at 60 percent of human communication. This includes our facial expression, body language, voice tone, pitch, flow or volume, words of emphasis or style of speaking."*

As she said, if we lose those mentioned features, *"could we still communicate?"* If we lose the awareness of culture, could we still communicate, and then, what would we share?

Very much alike, losing verbal, nonverbal, and the lens of culture, we lose a large part of our ability to communicate because, in fact, we lose most of the meanings we have to share.

At least in part, the loss of communication would generate frustration, stress, and mental implications.

It is known that communication derives from assembling features. We do not communicate a single word but an assemblage of them. Doing digitally or electronically, one uses a mathematically modulated language created for such a scope. This is a different language than the one produced by our minds for conversation.

As Tanya Leake observed on her students, the digital or cyber language *"have a negative effect on language itself, on cognitive, social and emotional development. Studies have demonstrated that our overuse of digital technology can result in overwork of the left brain and atrophy of the right brain and prefrontal cortex, resulting in symptoms of "digital dementia."*

In my opinion, the most drastic effect of digital technology will show a significant reduction in mental assembling that is caused by a decrease in language practice. It will significantly reduce our cognitive intelligence, while assembling technological elements will continue, producing an **ever increase in "intelligent" products**.

In nature, the evolution of phenomena depends on the scale the observation takes place. In fact, all of the phenomena show abrupt changes and transformations.

Especially the complexities collapse abruptly.

Our civilization evolves toward building a technological environment that develops independently of our "intelligence." The differentiation between these two increases exponentially.

All complexities evolve exponentially, but they collapse when their energetic need can no longer be fulfilled. There is no stagnation here because the evolution suddenly crosses an energetic threshold. It will not stop or stagnate on such a threshold. It will cross it while collapsing.

Sixth Paper
Homo loquens

Dan M. Mrejeru

The hypothesis of Homo loquens

Abstract

This paper aims to demonstrate, with hopefully robust arguments, that a unique chain of geophysical events (starting 43,000 years ago) influenced the Homo sapiens brain, transforming its mental setup, which resulted in a language-ready brain. This transformation led to a new species, which is distinct from the original Homo sapiens. I have called this new species Homo loquens because of its unique mental abilities based on language.

This research may be relevant for Anthropology, Linguistics, and Social Neuroscience fields while envisaging civilization's entire advent from a new perspective.

In my opinion, the language-ready brain, with its associated mental outcomes, are the only generators of the current civilization that turned out to be unrelated to any of the previous stone-age developments.
I have to mention that many of my papers uploaded on my site on *academia.edu* in the Social Neuroscience section, and the aspects presented here are interrelated. However, they provide plenty of details, argumentation, and references, which could not fit the current paper's format.

Humans became unique on this planet because of a language-ready brain and the further development of the language. The language-ready brain was the result of a complex transformation of the Homo neural networks. In the meantime, the information technology specialists showed that such language contains the textbook for generating all possible technologies.

The main factor that gradually changed the Homo brain was the repeated occurrence of geomagnetic events, which favored the cosmogenic radiation's penetration through the Earth's atmosphere, increasing by 20 to 80% the C 14 isotope atmospheric concentration.

The animal organisms assimilate C 14 Glucose when feeding on vegetables. In short, atmospheric C14 isotopes, as absorbed into C14 Glucose, stimulate oxygen species (ROS). And it regulates the production and inhibition of nitric oxides (NO), which both have interrelated roles in neurogenesis. This paper also aims to explain the differentiation in neurogenesis that occurred between Homo and other species.

The main study used in my investigation is *Low-dose or low-dose-rate ionizing radiation-induced bioeffects in animal models*, authored by Feng Ru Tang, Weng Keong Loke, and Boo Cheong Khoo of the University of Singapore. The paper was published in the Journal of Radiation Research (online 2016 December 27). (doi: 10.1093/jrr/rrw120). The authors said: "In this review paper, we aimed to update radiation researchers and radiologists on the current progress achieved in understanding the LDIR/LDRIR-induced bionegative and biopositive effects reported in the various animal models." The authors concluded: "*In summary, under certain circumstances, experimental animal data suggests that LDRIR/LDRRIR exposure may not only promote fertility and prolong the lifespan, but also induce immunological modification, give anti-tumor ability, slow progression of atherosclerosis, and ameliorate diabetic nephropathy. More data is needed to be generated to validate existing claims of biopositive/hormenic effects on humans.*"

The high prehistoric concentration of atmospheric C-14 isotope, assimilated in C-14 Glucose, transformed the neural networks and its architecture by increasing the neurogenesis processes that became solely directed toward a higher cognition capability and brain plasticity. Such a complex process generated a type of intelligence vastly different from that of Homo sapiens.

Introduction

When the animal's organism assimilates C14-Glucose, it inhibits nitric oxide production. The pharmaceutical experiments show that when a C14-Glucose is present, the slight trace of C-14 isotopes stimulates oxygen species (ROS) production while locally decreasing nitric oxide (NO).

However, during experiments, the C-14 traces have been between 1 to 5mCi., thus entering the low hazard category. By comparison, the average concentration of atmospheric C-14 isotopes at sea level is estimated to be 2mCi per year.

During the atomic bomb atmospheric testing, such C-14 concentration rose to 3.6mCi (80% higher than usual) as it was investigated in 1963-65, and it diminished to 2.4mCi in 1990 (20% higher than average). Only in 2020, the concentration returned to the preexisting level (1955).

Thus, the atmospheric data mentioned above could be correlated with the dose used in pharmaceutical and medical experiments. The effects of today's experiments can be compared with past and present natural occurrence of C-14 isotopes.

Fundamental differences would be produced by the length of the exposure time. In prehistory, many generations were exposed to high C14 isotope concentrations one after another for hundreds and thousands of years. This process ended around 4,500 years ago; a slight resurgence occurred again 2,700-2,400 years ago.

The C-14 isotopes inside the human body, as part of C14 Glucose, would stimulate the production of oxygen species (ROS), which produce oxidative processes. Such oxygen species have specific reactions with nitric oxide, regulating its concentration in various tissues. NO is known to create a nitrite reduction pathway. Within the blood vessels, the oxygen produces vasoconstriction that causes hyperbaric tension within the vessels. High tension is found inside the cells, inflicting the level of permeability of the cellular membranes. It increases the amount of available oxygen in blood vessels and cells and interacts with local nitric oxide, regulating its production.

NO, and ROS can be generated by the same enzyme or by different ones through alternative reduction and oxidation processes.

Nitric oxide is involved in transporting and dispersing Glucose in the entire organism. Glucose also stimulates oxygen species that metabolize any unwanted excess of this substance. When nitric oxide (antioxidant) decreases in the brain and the nervous system, it is so because local oxygen species increases. The oxygen species reduce the nitric oxide (antioxidant) locally, causing hyperbaric vascular tension by vascular constriction. By contrast, the same vascular pressure affects the vascular peripheries distinctly, making the nitric oxide increase locally (vasodilation) while stimulating muscle fibers and skeletal developments.

It must be said that ROS species can activate signaling pathways with conflicting and sometimes negative consequences, like reducing antioxidant capabilities for cardiovascular, renal, and central nervous systems.

On the positive side, ROS regulates cellular differentiation, proliferation, apoptosis, cell cycle, and migration.

The above effects are a reasonable explanation for the role played by physical exercising: they strengthen the muscles, but they also better oxygenate the brain, thus helping the cognition and memory processes. It is there a ROS and NO interplay stimulated by the oxygen produced during the physical exercises.

Method

I have researched Pub Med and other sources for English-language articles, mostly in peer-reviewed journals, to find the arguments needed to feed my hypothesis. I used the evidence provided by these sources.

I have to indicate that many similar or distinct sources have been subject to investigation in my previous articles on academia.edu.

Over time, my investigative research concluded on various issues I focused on, and the conclusions moved my analysis into new directions.

I started with the geomagnetic excursions; I analyzed their duration and radiative intensity. Unexpectedly, I found that the rate of occurrence for such excursions accelerated during the interval 43 ka-4.5 ka, while their radiative pulses intermittently covered a gap of 30,000 years.

The result of the mentioned radiative pulses led to a high atmospheric concentration of C14-isotopes. Then, I found that the atomic bombs' experiments in 1955-1963 have produced the same type of C14 isotope concentration as in our prehistory.

Searching medical literature, I analyzed the effects of various nuclear technology procedures and experimental health and pharmacological testing.

Of particular interest was a study made by a research group from the University of Singapore. This study (see reference) suggests that experimental animal data generates a host of biopositive effects; no humans have been tested yet in the same range of doses.

I found that medical procedures use a radiation exposure of 2-4mCi (millicuries) that produces a biological absorption of 0.4mGy/hr.

The same radiation exposure is recorded for the average C14 concentration at sea level 2mCi/year. When the C14 concentration increased in prehistory and during the atomic bomb experiments, the atmospheric C14 concentration increased by 20-80% (2.4mCi to 3.6mCi).

These findings made me think that I could interpolate the data on radiative effects and analyze the human brain's biological results in the last 40,000 years of our evolution. Besides a lack of direct information in the literature caused by a general scientific disinterest in this topic, I could still argue my hypothesis using a comparative investigation.

Materials

I like to mention here the study *Analysis of the atmospheric C14 record spanning the past 50,000 years derived from high-precision Th230, U234, U238, Pa 231, U 235, and C14 dates on fossil corals*, authored by a team led by Tzu-Chien, Richard G. Fairbanks, Li Cao, and Richard A. Mortlock and published in Elsevier (Quarterly Science Review 26) on June 19, 2006.

A second study I used on the same topic is *Atmospheric C14/C12 changes during the last glacial period from Hulu Cave (China)* authored by Hai Cheng, R. Lawrence Edwards, John Southon, Katsumi Matsumoto, and a long list of other colleges.

From the medical field, I used the study *Effects of Chronic Low-Dose Radiation on Human Neural Progenitor Cells* authored by a large group of Japanese and Chinese researchers led by Mari Katsura, Hiromasa Cyou-Nakamine, Qin Zen, and Yang Zen and published in Scientific Reports (article number 20027-2016) on January 22, 2016.

I followed this study with another one: "*Low-dose of low-dose-rate ionizing radiation-induced bioeffects in animal models,*" authored by a research team from Singapore University.

I have to mention the study *Reactive Oxygen Species in Metabolic and Inflammatory Signaling* by Steven J. Forrester, Daniel S. Kikuchi, Marina S. Hernandes, Qian Xu, and Kathy K. Griendling published PMC 2019 March 16.

Results

The geophysical studies mentioned above indicate that the concentration of atmospheric C14 isotopes was from 54ka to 50 ka, 12% higher than the levels recorded before 1955. Between 50ka and 43ka, it increased to 28%. From 43ka to 38ka, it was 60-80% higher. Here are 5,000-6,000 years of continual high C14 concentration with significant biological impact on the human brain.

From 38ka to 25ka, it was 60-40% higher. 20ka, it was 50% higher. From 23ka to 11ka, it was 50-60% higher.

From 10ka to 2.5ka, the value was 15% higher, but a peak was recorded 6ka as 30% higher.

In sum, during the last 30,000 years before the current era, the C14 concentration was 50-30% higher than before 1955.

When it is interpolated, this data indicates that the atmospheric C14 isotope concentration generated a higher radiation exposure than 3mCi that is 50% higher than an average recorded before 1955.

At its peak (1963-1965), the radiation in the atmosphere reached 3.6mCi, identical with the geophysical records for the era 43-37ka.

The medical study mentioned above referred to experiments with three types of low-radiation-doses on human brain biology. The lowest dose used was 31mGy, representing a 4mCi radiation exposure that produces an absorption of 0.4mGy/hr. The effects of such 31mGy exposure were experimentally proved to have no statistical significance.

In the meantime, this last dose still produced an increase of 1.5-fold in the gene expression, affecting 6% of all gene loci.

The study found that "*low doses of radiation in the upper range of common diagnostic procedures create mutations through inserted DNA even more efficient than the much larger doses.*"

These experimental medical findings suggest that atmospheric high C14 concentrations of 50%-80% above normal could produce significant increases in gene expression and could trigger higher than regular rates of mutations. This was a situation that persisted for almost 30,000 years during our recent prehistory.

The other study mentioned above was about the experimental evidence obtained from animal models. They exposed animal models for a prolonged time during testing, like several tens of generations (20-30 generations).

The mice were exposed in one test to a constant low-dose of 4.3mGy/day for three weeks. In another experiment, a mice colony was exposed for 21 generations to a 28.8mGy dose at 1.2mGy/h.

Many biopositive effects were recorded in both cases, like increased litter size, more fertile than the control group, increased litter number, increased viability, and faster growth rate.

However, when the radiation doses were higher than 30mGy, the researchers observed the first but mild, bionegative results.

Discussion on the role of oxidative stress

A chemical reaction in cells caused by oxidation produces free radicals, which are the primary source of biophotons; this process occurs during the deexcitation of free-radicals.

All biological systems have a safety mechanism that rapidly intervenes to deexcite the most dangerous free-radicals.

As for brain plasticity and its connection to oxidative stress, I will first define plasticity. It refers to an undifferentiated cellular state, where the cells, not being differentiated, are open to all options. Consequently, they are available to adapt to needed changes.

Neurogenesis generates undifferentiated neuronal networks. When the entropy occasionally increases in the hippocampus's Dentate Gyrus, a neuronal generation will burst with higher than average plasticity (more undifferentiated neurons).

It is known that higher oxidative stress generates a higher entropy. Hence, an increase in cellular entropy is directly produced by the rise in oxidative stress.

It is well established that short pulses of augmented entropy/stress result from oxygenation pulses (oxidative stress). The process produces short pulses of free-radicals, which are almost immediately annihilated by the deexcitation. The result is biophoton production.

Scientific literature estimates that biophotons would play a significant role in neuronal transmission and communication. A high biophoton neural production was part of the change in the Homo new intellect.

Discussion on nitric oxide neurotransmitter functions

When the neurotransmitters are involved, a higher or lower amount of nitric oxide will diminish or favor communication between them and the neurons. The nitric oxide synthase plays a catalytic role that makes neurons resistant to toxic insults and neurodegenerative disorders.

The research shows that it is involved here in the neurotransmitter of the *cerebral vasodilator nerves*. In cases the nitric oxide synthase is produced in excess, it functions as a *neurotoxin* that causes neurodegenerative disorders.

Nitric oxide (NO) acts as a neurotransmitter, but it is synthesized only on demand. It diffuses with protein receptors of the adjacent cells. This is why NO is regarded as a signaling molecule and a secondary neural messenger.

Because it generates a reduction in the nitric oxide at the neural level, this process's effect is vasoconstriction. As it is known, vasoconstriction produces an effect similar to the administration of hyperbaric oxygen.

Here, we can compare the hyperbaric oxygen treatment results with the effects produced by the prehistoric high atmospheric concentration of C-14 isotopes.

The oxygen species (ROS), being stimulated during neurogenesis from a hyperbaric effect, reduces the nitric oxide's neurotoxin outcome.

However, it seems that when the nitric oxide is diminished in the brain, in compensation, it increases somewhere else, where it becomes a stimulant in metabolism, in skeletal and muscle contractility.

Discussion on the shaping of the new brain: the combined oxidative vasoconstricting effects with nitric oxide vasodilatory effects

The combined effects of enhanced neurogenesis result from a stimulant C-14 Glucose and associated oxygen species production at the neuronal level. The compensatory effect would help a human significantly improve physical abilities by developing muscles and gaining stronger bones.

In the meantime, reducing the nitric oxide at the cellular level will increase the vasoconstriction effect, thus supplying more oxygen to the cellular functions. A diminished biological exposure of migrants to new pathogens and viruses encountered on the path out-of-Africa resulted from a sufficient cellular oxygenation increase.

All the aspects mentioned above indicate a sum of combined favorable features being developed on the path out-of-Africa. It would help the migrants improve their mental condition, associating it with better physical performance, in the tentative to overcome each local while novel challenge.

When the Laschamp excursion occurred 42kyr, Homo sapiens were leaving the tropical and subtropical zones of Eurasia, marching through the temperate areas. It corresponded too to a temporary warming up of the climate.

When the Last Glacial Maximum episode started 25 kyr, Homo sapiens were already on the way to gain a radically improved language. They achieved tremendously novel mental capacities and a more assertive physical posture. These features allowed Homo to vastly improve their ability (brain plasticity and cognition) to adapt to harsh and rapidly changing environmental challenges, like crossing high mountains and entering frigid zones. No such abilities had existed during Homo's existence in Africa.

I would say that the above-mentioned new abilities positively contributed to the success of out-of-Africa migration.

Thus, those changes occurred within Homo's mind by higher cognition and plasticity and acellular adaptation to novel Eurasian pathogens and viruses. Such changes in Homo around 40-30 kyr were instrumental in propelling the migration further into the temperate, subarctic, and even Eurasia's arctic zones.

The LGM (Last Glacial Maximum) temporarily transformed many temperate and mountainous zone toward a frigid climate.

However, it took another 10,000 years until Homo was mentally prepared to understand how to build new stone tools (like the Gravetian/Magdalian types), earliest petroglyphs, earliest archery tools. But 30kyr+, the first rock art, appeared in Europe and Borneo.

I would say that those 6,000-7,000 years of LGM determined the incipient novel capabilities to switch toward a cold adaptation as a temporary distraction.

The starting of deglaciation around 18 kyr corresponds with a boom in every possible approach: from evolved tools, like axes and bows, to first open-air temporary shelters to animal and plant domestication. The social development induced a self-domestication that functioned as an intermediary for further social development.

Who was in charge there: Homo sapiens or Homo loquens?

In my opinion, Homo sapiens ceased to exist around 20 kyr.

Globally, all that remained was only one species: Homo loquens.

Discussion on competing roles of oxygen and nitric oxide.

As observed during various experiments on plants and animals, a reduction in the released nitric oxide causes vasoconstriction.

As a team of researchers from Nanjing University (China) demonstrated in their paper "*Bidirectional Regulation of Neurogenesis by Neuronal Nitric Oxide Synthase Derived from Neurons and Neural Stem Cells*" (published in Stem Cells in 2010), "the neuronal nitric oxide synthase (nNOS) negatively regulates adult neurogenesis."

Another paper ("*Nitric oxide negatively regulates mammalian adult neurogenesis*") of a team of researchers from Cold Spring Harbor Laboratory, NY (published August 5, 2003, in PNAS) arrives at the same conclusion. The authors maintain that:

"Here, we report that nitric oxide (NO) acts as a negative regulator of cell proliferation in the adult mammalian brain."

The hyperbaric oxygen, through vasoconstriction, increases the tension inside the blood vessels (hyperbaric effect).

Two other papers ("*Long Course Hyperbaric Oxygen Stimulates Neurogenesis and Attenuates Inflammation after Ischemic Stroke*," written by a team of researchers from two Taiwan universities, and "*Hyperbaric oxygen therapy promotes neurogenesis*," written by a group of researchers from Loma Linda University, US, and Chongqing Medical University, China) demonstrate how hyperbaric procedures stimulate neurogenesis.

Discussion on uniquely developed human neurogenesis.

As the literature indicates, humans present a different pattern of adult hippocampal neurogenesis than other mammalians. In humans, the vast majority (90%) of neurons in the dentate gyrus are subject to exchange, compared to 10-30% in other mammalians investigated. Humans show a less pronounced age-dependent decline during adulthood compared to mice and other mammalians.

A little addition of new neurons to the olfactory bulb (OB) after the perinatal period seems unique to humans. As various studies show, in other mammals, the neurons generated in the subventricular zone are integrated preponderantly in OB (olfactory bulb) during their entire life cycle. But in humans, after their birth, there is not such a process. It could be said that the rate of addition of new neurons to the Olfactory Bulb is meager if compared to the one to the hippocampus. It generates a fundamental difference in the human brain's plasticity, which differentiates humans from animals. It details how human neurogenesis takes a different direction from one of most mammals.

The contribution made by neurogenesis to human cognition/plasticity determined our brains to switch from olfactory determination to context appreciation. Language gradually enriched all perceptions with a context. It led to correlative thinking based on contextuality.

The same increased plasticity produced a new mental foundation that allows the brain to operate with the symbols included in the words developed within the language. It generated that mechanism that defines the language-ready brain.

However, one of the most critical processes followed was the rapid switching from sign languages to verbal language utterances. While the sign languages used the left hemisphere, most of the other features made our prehistoric brain display a right hemisphere preponderance and dominance.

Language development implied a switching of the correlates that moved the dominance to the left hemisphere.

Discussion on Enantiodromia

Carl Jung introduced it as a feature about the emergence over time of an unconscious opposite. It is saying that any extreme is opposed by a tendency to restore balance within the natural world.

Recent studies found that atypical individuals who show unilateral language dominance in the right hemisphere can score high levels of complex thinking. Such an architecture can support exceptional instances of intuitive insight in any problem-solving.

Such scientific evidence is a strong indication of the previous dominant role of the right hemisphere. One could say that the old dominance was based on an extensive, unconscious information processing (uncertainty) that was dealing with much more information input than the brain currently operates. Such input was nonlinear, and it needed conscious conversion. That brain needed extensive neuronal circuits and a lot of energy consumption. It may also explain the large Paleolithic brains, which started to diminish in the last 12,000-10,000 years while reaching today 10% less volume.

It is proof that self-domestication (or the friendlier survival) results in smaller brains among mammals. This is true, especially in humans.

These arguments may lead to a hypothesis that the development of the left hemisphere, as the region that deals with language, occurred as the emergence of a natural counterposition to a too busy to coop the right hemisphere.

There is here a question on the switching of correlates: why does such a distinction arise? A simple answer would be: it was a natural way to balance brain activity.
A second answer would be that evolution is not linear despite our simple way of seeing it.

Discussion of the role played by many species and subspecies of hominins.
In the humans' case, we have to consider an entire range of species and subspecies originating in monkeys as our evolutionary connections. I would think of a nonlinear view, where the fundamental argument is related to neurogenesis.

As one could observe, the way neurogenesis evolved in mammals and primates points to a change when a clear separation singled out the humans against the rest. If one counts the subspecies and species that have evolved out of the Australopithecus in the last 4 million years, one can find many species and subspecies. I would say that archaeology probably could not find more than a few of all the species which had existed in this large family of so-called hominins.

Why has such a vast family of hominins developed? The first probability to consider refers to the high-sensibility of this hominin family to environmental changes. The second probability to the many varied or distinct species and subspecies from where it would be chosen natural optimality.

Discussion on geophysical and radiative events producing biological causality.

A specific causality that took advantage of an existing biological sensitivity within a nonlinear perspective should be considered. As mentioned, the humanoid neurogenesis was affected by a particular geomagnetic occurrence that is a geomagnetic excursion. Here is a list of the main geomagnetic excursions of the last 300,000 years:

-Calabrian Ridge 260 kyr

-Pringle Falls 211 kyr

-Iceland Basin 188 kyr

-Blake 120 kyr

-Norwegian-Greenland Sea 80-70kyr

-Laschamp 41kyr

-Mono Lake 33-31 kyr.

-Lake Mungo 29-26 kyr

The geomagnetic field was in excursional mode for the entire interval 31-23 kyr with the lowest magnetic intensity developed 37 kyr.

After deglaciation, the excursional mode continued with four-six more events. Geophysics recognized only four of them:

-Hilina Pali 18 kyr

-Gothenburg 13.5-12.5 kyr

-Solovki 7.5-4.5 kyr

-Sterno-Etrussia 2.8-2.2 kyr.

One can observe that the densest radiative events took place between 41 kyr and 4.5 kyr. The events were intermittent but lasted for hundreds to thousands of years on each occurrence.

This was when the Homo sapiens brain was subject to forced modeling, causing its fundamental transformation that generated the language-ready brain.

Discussion on the biological clue of nitric oxide.

As it seems, the agent of such complex but somehow contradictory evolution is nitric oxide (NO). It is a gaseous molecule synthesized by the enzyme nitric oxide synthase (NOS). It acts as a neurotransmitter. It is also a part of the signaling pathways that operate within the cerebral vessels, neurons, and glial cells.

NOS is known to exist in three isoforms: endothelial nitric oxide synthase (eNOS), neuronal nitric oxide synthase (nNOS), and inducible nitric oxide synthase (iNOS).

These isoforms influence the brain in distinct ways because they are active in different regions within the brain. The NO signaling pathways regulate the cerebral blood flow (CBF) during the rest and various stresses.

In the meantime, there is a growing body of evidence that NO plays a significant role in numerous biological processes. It is an essential mediator in the central nervous system, and it is supposed to be implicated in various neurological diseases. For example, in the degenerative brain processes, which affect the CNS, the nNOS level rises rapidly. In some pathological conditions, as inflammations, the iNOS level is very high.

Overall, the nitric oxide seems to act as a "double-edge-sword." The role of nitric oxide could contribute to either cell survival or cell death. The ultimate answer is the untimely transfer of nitric oxide from one protein to another. There is evidence that NO transfer could lead to cellular suicide.

NO can act as a modulator of the inhibitory processes within the synaptic transmission. But also, it can modulate the activity of ion channels and regulate the ion exchanges.

The C14-Glucose, ingested through hominin and human feeding, is known to have, in general, an inhibitory role on nitric oxide.

Discussion on the mental transformation of Homo sapiens into a novel species.

Now, I like to refer to Homo sapiens evolution. As it is known, it evolved from an "archaic type" (around 190ka) into an "incipient sapiens" (120ka). Nevertheless, around 80ka-70ka, the first manifestation of art in South Africa appeared. It is true also that Neanderthals seem to enter a phase of high development during the same epoch. But around 42ka, a chain of extreme geophysical events started to unfold. This chain generated a revolution in the primitive language of Homo sapiens. Still, it led to the extinction of Neanderthals and

Denisovans, while the last surviving Homo erectus and other hominins also reached their extinction phase. It is a critical point, where Homo sapiens, in a relatively short time, encountered fundamental changes, like the change in skull geometry toward rounding, the conjugates switched the brain lateralization, dual-use of the brain neural circuits (visual + verbal), some new neural circuits (for verbal use only), the changes in the neuronal processing speed (the speed diminishes). Simultaneously, the entire body suffered a process of gracilization (as an extended form of neoteny). The change in the skull geometry is seen as fundamental for a very distinct development of language. It had put the language-related sections in a central position, and by rounding, it diminishes some circuitry distances. A switching of correlates had paralleled the other developments in the brain while changing the pattern of lateralization. As the archaeology indicates, the skull's rounding appeared some 30ka but was fully completed only during the Holocene optimum (cca. 8ka).

According to many specialists, consciousness itself settled in a form similar to the present only 4,500 years ago.

The language was generating anatomical changes because it was a technology that became embedded in our biology. It was a complete biological response that needed to be developed by this new species: from genetic changes to new genetic expressions to anatomical reshaping.

One argument in favor of the language's mechanical role is switching the correlates from the right hemisphere to the left one, which changed the hemispheric dominance.

Discussion on language as a technology that stimulates our imaginary approaches

The language was and is a technology in itself. It can infinitely produce other technologies by playing with its assembling/disassembling capacity.

Time and again, the language came with a set of features, where communication was one of them but not the most important. The other elements were planning, control, and socialization. Over time, the importance of each feature changed. Planning implies creating careful coordination of actions. Hence, planning generates, among other things, all other languages, including those we call algorithms. We can infinitely create languages and algorithms.

Control confines us to a system of minimal variability that excludes the rest as unperceivable and unthinkable. Over time, we have learned how to deal a little bit better with variability. But even then, our progress was minimal.

Socialization seems to be one of the best-developed functions of the language.

Discussion on the visual vs. the verbal brain

A recent study at the Psychology Department of Harvard Medical School, authored by Elinor Amit and Evelina Fedorenko (published on May 11, 2017, in The Harvard Gazette and journal NeuroImage), has highlighted several necessary details of modern thinking. The authors have demonstrated that people use inherited visual thinking about the things which are closer to them. On the other hand, people use inner speech as their new verbal ability when contemplating far-off things.

In short, people use visual thinking to deal with self, in-group, past, present, and near-present. By contrast, verbal thought is used when dealing with somebody else, an outgroup, or the future. The researchers found out that the visual deals with the past and the present, while the verbal agrees with the future. Here is a clear indication that verbal thinking development helped us define the concept of the future.

More importantly, it looks like language helped individuals deal with each other while ultimately defining and promoting each one.

It also helped the individual to become open to out-of-group relationships and contacts. A less talked about aspect is the relationship that started to develop with the environment. Such a new approach turned paramount in human evolution: it led to animals' and plants' domestication.

In short, domestication changed the type of social relationships, but this domestication has led to a change in diet and the kind of activity. Everything has been promoted thanks to the language.

In the meantime, the researchers found out that, at present, there is no pure visual or pure verbal thinking because both of them are mixed in every expression. Hence, the recently developed verbal mode is, by now, very well implemented in all of our brain circuits, contributing to a specific part of every expression. Even so, one of these two modes of thinking still appears to be predominant in one expression or another.

As a negative outcome, experimentally, it is demonstrated that those people very good at verbal thinking score about average, or even low, on the object and spatial visualization tests.

Almost all the old visual circuits have been affected by the interference with the newly developed verbal mode. This is an interesting mental approach because each of these two modes shows a very different processing speed (it is estimated that the visual one is at least 10,000 times faster than the verbal one).

Overall, this Harvard study provides valuable information about how human thinking has radically changed since the advent of language, starting 30,000 years ago.

I see that the most significant result of verbal was the development of "*imaginary thinking*" accompanied by its "*factual*" products. This "*imaginary*" mode exploited the verbal capabilities to a maximal extent that implies an "*infinity*" of mental manipulations, which generate "factual" products.

However, the language resulted from an intricate play of evolution that adapted the hominin and human biology to particular geophysical constraints.

Discussion: was language an antidote to extinction or only offered a transition path for a change to another species?

Did language help us avoid extinction? It is known, all other hominin species disappeared at the same time when humans began to turn into a new species, cca. 30ka. Hence, Homo sapiens disappeared too because it was replaced by Homo loquens in no more than 2,000-5,000 years.

In support of my opinion, I would like to add a controversial idea that still makes sense. It refers to all cultural descriptions that picture our origin.

I would focus only on the Bible, where the symbolism is much clearer. The Bible says: "In the beginning was the *word*." Then, I am wondering: who spoke those *words*? The Bible makes no mentioning of any "primordial human" who would make stone tools and resembles Homo sapiens.

Thus, in my opinion, the Bible characterizes the first humans as speaking beings, and in doing so, it defines humans as a new species that is distinct while disconnected at the cerebral level from any other species inhabiting the Earth.

On the other hand, the biblical description, as it seems, pictures the "*word*" as an almighty feature or as the most critical human quality that helps "*tooling everything else*." But no other "tool" is mentioned there because the "*language is a tool generator in itself.*"

Conclusion

This paper aims to prove that the radiative event that contributed to the environmental presence of C-14 Glucose had a complex biological effect that stimulated a type of enhanced neurogenesis in the Dentate Gyrus. As a process that proliferates and differentiates the migrating neural precursor cells, neurogenesis reflects neuronal intrinsic (genetic) and environmental influences (like C-14 isotopes).

The adult-born neurons in the Dentate Gyrus of the hippocampus exhibit critical periods of long-term plasticity during their maturation, contributing to the development of the mechanism of learning and memory. More importantly, new neurons expand the plasticity capacity while enhancing the pattern separation processes that enlighten the analysis/synthesis mechanism that copies the language's assembling/disassembling tool.

Language occurred due to a forcing geophysical factor that was cosmogenic. The universe imposed a high-rank alternative that moved us into a higher hierarchy.

However, because of language, humans built everything imaginarily (in a symbolic manner) in contrast with the natural reality.

I would say that the transformation was a repeating process, intermittently stimulated, which gradually increased the overall effect.

Thus, while the transformation was rapid into a mental fundamental, it needed time to produce interrelated connections and inner neural network adaptations. The processes fed one another, making the new cognitive skills evolve gradually, accumulatively, and progressively.

In the end, it may be a novel argument provided by very recent genetics studies.

As a paper pre-published on BioRxiv (doi: https://doi.org/10.1101867317) on December 6, 2019, indicates that "*analysis of ancient and present-day non-African chromosomes, all points to East/South-east Asia as the origin 50,000-55,000 years ago of all known non-African male lineages. This implies that the initial Y lineage in populations between Africa and eastern Asia have been entirely replaced by lineages from the east, contrasting with the expectations of the serial-founder model, and thus informing and constraining models of the initial expansion*".

"*No ancient Y-chromosomal data earlier than 45,000 years ago have been reported, but 21 Asian or European males living 30,000-45,000 years ago are documented, and for 18 of them, assignments to C, D, or FT have been reported. First, none of the ancient samples carry Y lineage outside the 23 represented in Figure 1 at 50,000 years ago. Second, C lineage, now confined to East, Southeast and South Asia plus Oceania, were more widespread 30,000-40,000 years ago, including in Europe where they persisted until after 8,000 years ago*".

In my opinion, the mentioned study, along with several others, shows that the initial individuals who migrated out-of-Africa some 55,000-50,000 years ago were utterly replaced around 42,000 years ago by other novel individuals, who originated not in Africa but within South and Southeast Asia.

I have to say that probably during a better recalibration of data, the said 45,000 years old samples would test to be only 42,000-41,000 years old.

This genetic research seems to favor my opinion that Homo sapiens was gradually replaced by a new species with a distinct "intelligence" (Homo loquens).

Another very recent research on dating the last samples from the Neanderthal skeletons also indicates a much abrupt disappearance of the Neanderthals around 42,000-38,000 years ago.

This means that Neanderthals and Homo sapiens disappeared in the same interval and connected to the Laschamp Geomagnetic event.

Acknowledgment

Because of this paper's limited length, I was forced to diminish the amount of cited and quoted sources to a restricted number. Thus, some of the information provided may appear superficially treated. It is based on a vast amount of scientific data. Those details are in my self-cited work uploaded on academia.edu

References

Tzu-Chien, Richard G. Fairbanks, Li Cao, and Richard A. Mortlock, study *Analysis of the atmospheric C14 record spanning the past 50,000 years derived from high-precision Th 230, U234, U238, Pa 231, U235, and C14 dates on fossil corals*, published by Elsevier (Quarterly Science Review 26) on June 19, 2006.

Hai Cheng, R. Lawrence Edwards, John Southon, Katsumi Matsumado study *Atmospheric C14/C12 changes during the last glacial period from Hulu Cave (China)*.

A research team led by Mari Katsura, Hiromasa Cyou-Nakamine, Qin Zen, and Yang Zen studied the *Effects of Chronic Low-Dose Radiation on Human Neural Progenitor Cells*, published in Scientific Reports (article number 20027-2016) on January 22, 2016.

Feng Ru Tang, Weng Keong Loke, and Boo Cheong Khoo *authored Low-dose or low-dose-rate ionizing radiation-induced bioeffects in animal models,* published in the Journal of Radiation Research (2017 Mar. 58: 165-182.

Steven J. Forrester, Daniel S. Kikuchi, Marina S. Hernandes, Qian Xu, and Kathy K. Griendling, a study *Reactive Oxygen Species in Metabolic and Inflammatory Signaling*, published in PMC 2019 March 16.

Chun-Xia Luo, Xing Jin, Chang-Chun Cao, Ming-Mei Zhu, Bin Wang, Lei Chang, Qi-Gang Zhou, Hai-yin Wu, Dong-Ya Zhu, *Bidirectional regulation of neurogenesis by neuronal nitric oxide synthase derived from neurons and neural stem* cells, Stem Cells 2010 Nov., 28 (11); Prehistoric 2041-52 10.1002/stem.522

Michael A. Packer, Yuri Stasiv, Abdellatif Benraiss, Eva Chmielnicki, Alexander Grinberg, Heiner Westphal, Steven A. Goldman, and Grigori Enikolopov, *Nitric oxide negatively regulates mammalian adult neurogenesis*, published by PNAS August 5, 2003, 100 (16) 9566-9571

Ying-Sheng Lee, Chung-Ching Chio, Ching-Ping Chang, Liang-Chao Wang, Po-Min Chiang, Kuo-Chi Niu, and Kuen-Jer Tsai, *Long Course Hyperbaric Oxygen Stimulates Neurogenesis and Attenuates Inflammation after Ischemic Stroke*. Published in Macrophage-Mediated Inflammatory Disorder, Volume 2013, Article ID 512978. Posted on February 21, 2013.

Jun Mu, Paul R. Krafft, and John H. Zhang. *Hyperbaric oxygen therapy promotes neurogenesis: where do we stand?* Published in Medical Gas Research online 2011 June 27

Peter Reuell, Harvard Staff Writer. *The power of picturing thoughts*, published in Harvard Gazette on May 11, 2017

Dan Mrejeru-*Cosmogenic factor*, published in August 2015 in Solovki Ersatz by Authorhouse.

Dan Mrejeru-*Prehistoric high concentration of atmospheric C-14 isotopes* (part 1 and 2), academia.edu, 2020

Dan Mrejeru-*Changes in brain anatomy have inflicted the development of language*, academia.edu 2019

Dan Mrejeru-*The causes that produced the development of language*, academia.edu 2020

Dan Mrejeru-*Absorption of C-14 isotope stimulates the neurogenesis and biophoton production*, academia.edu, 2019

Dan Mrejeru- *Conjugates switched the brain lateralization and generated a new type of intelligence*, academia.edu 2019

Dan Mrejeru-*Universal technology of language*, academia.edu, 2019

Dan Mrejeru-*Atmospheric C-14 shaped the modern brain*, academia.edu, 2019

Function

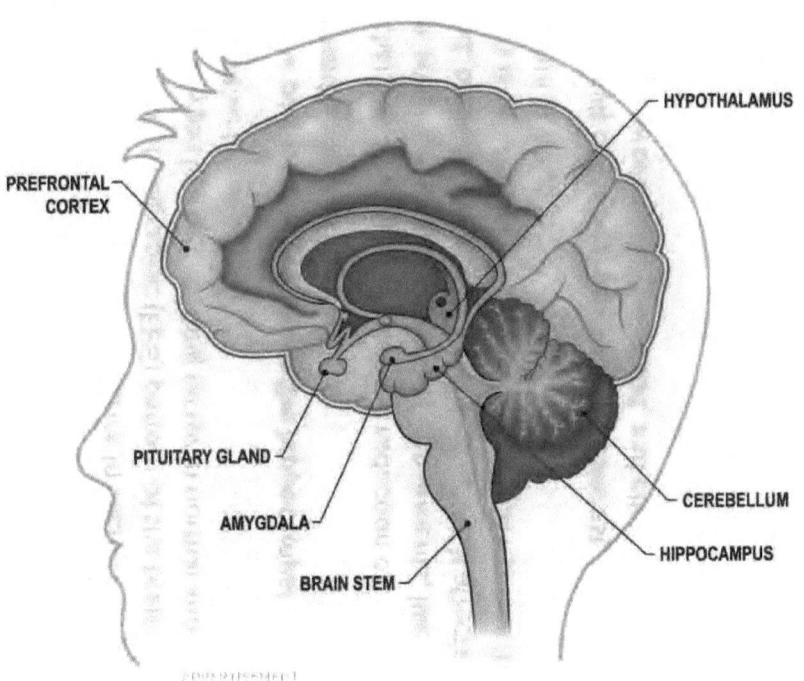

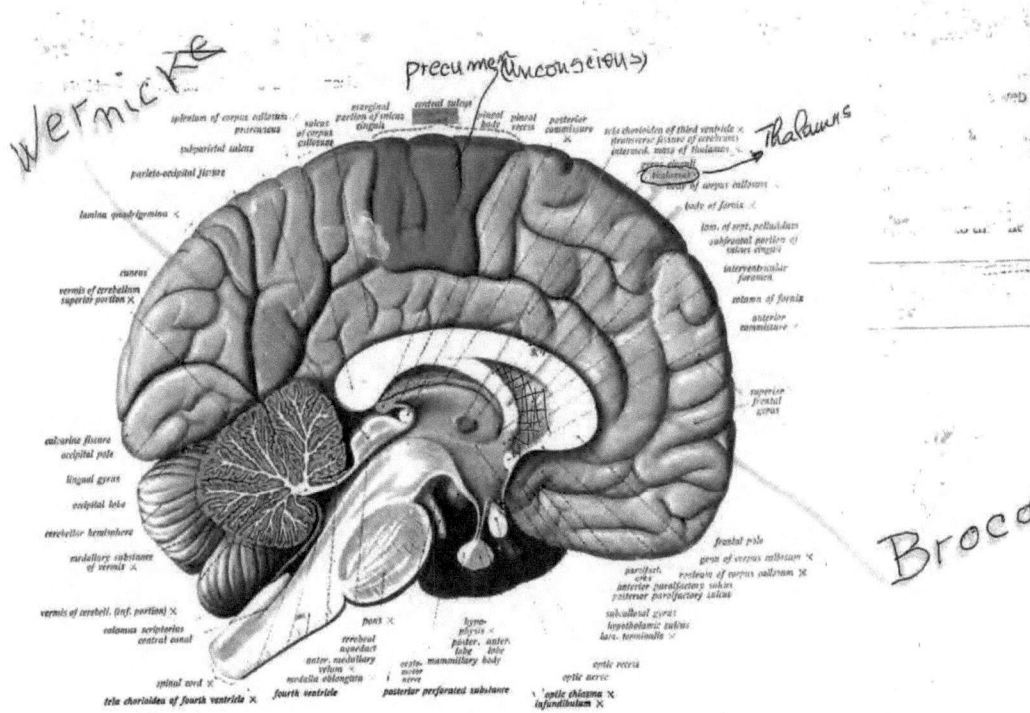

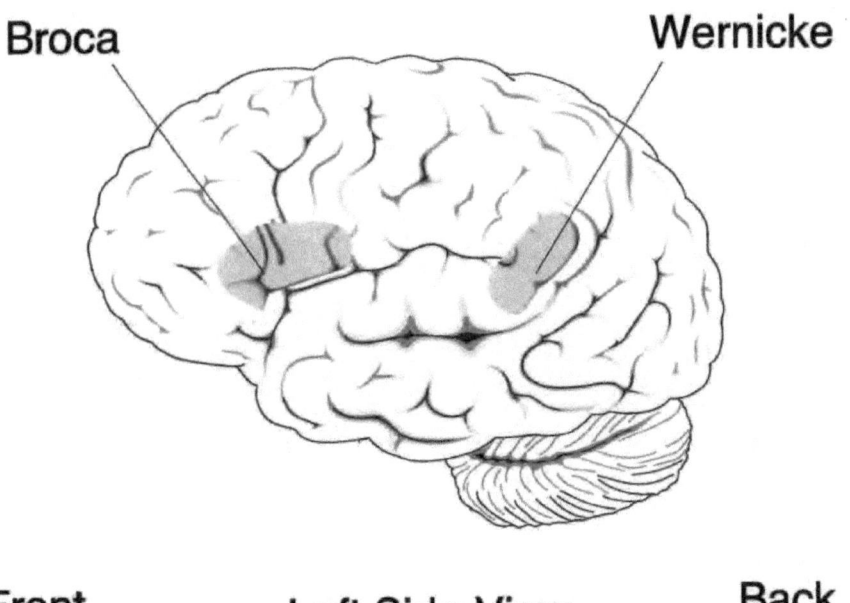

5th International Conference on New Findings on Humanities and Social Sciences

Certificate of Virtual Presentation

Presented to

Dan M. Mrejeru

Who participated in the 5th International Conference on New Findings on Humanities and Social Sciences and presented the paper
"The hypothesis of Homo loquens"

Farzam Chakherlouy
Chairman of Organizing Committee

20 - 22 November, 2020
Paris - France

www.hsconf.org

HSCONF-7058

Seventh Paper
Challenges

Neural challenges of the gaming brain

What would imply for our brain if humanity would manage to transform verbal language into an optic language? I need to define how exactly this tendency manifests and how it would evolve. For this reason, I would extensively quote several reviewing papers.

Our brain would have to use high-energy consumption over the current peak. Assumedly, current verbal mode introduced within the common visual pathways the low verbal speed that is said to be 20,000 times slower than visual mode. It is also said that the brain converted its energy needs by adapting to slow learning mode, saving a good part of its energy consumption.

If the brain volume, as gray and white matter, **would increase**, we would need a bigger skull. As most specialists would agree, the possibility of optic language cannot be accommodated by our cerebral anatomy in a relatively short time.

Another issue is the current **brain lateralization** set on the left hemisphere's dominance of the primary verbal mode. The recently established dominance already represents switching from a prior right hemisphere that accommodated the singular visual way of most of the Paleolithic era.

As anthropologists estimate, the language revolution manifested 30,000 years ago. It continued to enlarge the human skull until 25,000-20,000 years ago. Then, a gradual adaptation of visual pathways to verbal mode signified a new trend that started to reduce the volume of the brain.

It is known that white matter consumes most of the brain's energy. The reduction process probably affected the white matter architecture and connectivity, and this process took some 17,000-10,000 years to reach the current volume.

* * *

The first discussion implies a determination on **which segments of the brain participate in energy consumption**.

According to the study *Appraising the brain's energy budget* by Marcus E. Raichle and Debra A. Gusnard, published in PNAS 2002 Jul 29, the brain accounts for about 20% of the oxygen and, hence, calories consumed by the body.

In their study, the authors interpret the data provided by two studies of Yale University published in the same issue of PNAS and combine magnetic resonance spectroscopy (MRS) with an extracellular recording of neuronal activity in the cerebral cortex of the anesthetized rat.

In this study, both paths of investigation converged on the "*metabolic requirements associated with glutamate signaling in the brain. This focus would seem reasonable, considering that greater than 80% of neurons are excitatory and greater than 90% of synapses release glutamate.*"

"*Estimates from their approach (Attwell and Laughlin) indicate that most of the energy used in the brain is required for propagation of action potentials and for restoring synaptic ion fluxes after receptors have been stimulated by the neurotransmitter. In contrast, maintenance of the resting potential in neurons and glial cells accounts for less than 15% of the total energy consumption. Shulman and his colleagues, in a very different approach using MRS in anesthetized rats, have shown remarkably converging evidence that a large fraction (80%) of the energy used in the brain is correlated with glutamate cycling and, hence, active signaling processes.*"

There is "*an almost difference 4-fold between gray and white matter in both oxygen consumption and blood flow.*"

This study may give an idea of which brain systems were subject to a volumetric reduction in the last 20,000 years. The current reduction is estimated to be 15% to 20%.

Julia J. Harris and David Attwell, in the study *The Energetics of CNS White Matter* (published J Neurosci, 2012 Jan 4; 32(1):356-371), show the following figure.

"*We show that the white matter synapses consume less than 0.5% of the energy of gray matter synapses, and that is the main reason why CNS white matter uses less energy than gray matter.*"

"*This is surprising because: the white matter is half of the human brain; plasticity of the white matter is increasingly being invoked as a mechanism of learning; a disrupted energy supply to the white matter can cause vascular dementia.*"

*   *   *

I will continue the discussion on **metabolic/energetic functioning**, quoting the first authors cited above (Marcus R. Raichle and Debra A. Gusnard).

"*We might therefore posit that, in the brain, a large majority of its metabolic activity is devoted to ongoing synaptic processes associated with maintaining a proper balance between excitatory and inhibitory activity.*"

"*In part, it would seem to place emphasis on transient metabolic changes associated with alterations in the correlational structure of the neural circuit.*"

It is often said that 60% of the brain is involved in vision. But here, less than 20% is dedicated to vision-only because the rest of 40% is related to vision-touch, vision-motor, vision-attention, vision spatial navigation, vision-meaning, and many other combined functions.

The vision-only seems to reside in the V1 area, but this area was found to participate in motor behavior, too. The specialists consider that V1 is responsible for 10% of vision processing.

Other brain statistics indicate that 30% of the cortex processes vision, 8% olfaction, 5% touch, and 3% hearing.

The surveys indicate that 65% to 80% of the people are visual learners.

However, at least 60% of the white matter is dedicated to visual pathways, which have lengths of 5 cm to 10 cm and thicknesses up to 2 cm.

* * *

The next aspect that must be determined is **which neural structure processes the visual and which one is in charge of verbal processing**.

Thirty thousand years ago, when the language revolution occurred and the language-ready brain was anatomically formed. There were no well-defined language-ready pathways. Hence, gradually, the brain adapted the existing visual pathways to a combined mode of visual-verbal processing. Today, the researchers could not identify any exclusive verbal pathways.

As most neuroscientists would agree, all brain pathways have multiple functions (even if one seems dominant) because of fundamental plasticity that allows adaptation.

The organizational changes in the brain anatomy reduced the length of pathways. It allowed the opticoders, those adapted to carry sound signals, to spend a reasonable amount of time between structures.

* * *

Now, I like to discuss the **challenges that allowed the brain to process the sound produced by language**.

It is known that visual processing is assured by opticoders, while verbal/sound processing relies on lexicoders. Hence, the brain before language predominantly used opticoders, while the sound was processed in exclusivity for hearing. Thus, no stricto-senso lexicoders existed.

Probably 70,000 years ago, an incipient change in brain architecture allowed the opticoders to adapt to an increased amount of vocalization manifested as part of the sign language. Some opticoders assumed a new executive role, becoming lexicoders. They helped process the vocal expressions, gradually assembling vocal structures within a language.

The linear space of conscious perception results from a <u>mental hyperplane dimensional adjustment</u>. The rest of reality (over 95%), which is not affected by such a mental filter, is processed at an unconscious level.

Some opticoders would process the visual signal before the hyperplane action. I call these coders **A-opticoders**. They provide information to the unconscious.

Other opticoders have an adaptation (due to hyperplane reduction) to provide a linear interpretation of the visual signal. I will call these coders **B-opticoders**. They were better suited to deal with linear sound signals.

However, some coders existed in the brain to process sound for hearing needs. Assumable, these sound coders were implicated in dealing with the vocalizations of sign language and later to language processing.

The hearing processing of sound works as a filter that determines the recognition of the sound patterns. This filter may disassemble the sound elements to segregate particular patterns. But it does not allow assembling of the resulting structures of the sound patterns.

The hearing system does only a comparative analysis.

The brain used visual and sound processing on the same pathways.

Another possibility would be that the lexicoders originated in B-opticoders. They were also adapted to a dual-use of the pathways.

It took a while until the lexicoders succeeded in making a complete and efficient adaptation. This is in line with a relatively slow language evolution that followed up the initial revolution.

In the end, the complete adaptation of the newly formed lexicoders allowed the assembling/disassembling/reassembling of meanings and patterns. Such an assembly process was not suitable to prior opticoders because of their nonlinear functioning.

Even the B-opticoders linear processing cannot allow an assemblage like for the linear structures of the language.

Before the language, the Paleolithic man dealt primarily with unconscious thoughts based on visual/sound processing, which had a limited conscious impact, and produced a primitive consciousness.

* * *

Let's discuss the **advantages of unconscious processing**.

However, the unconscious thought had outstanding profitability that helped a specific efficiency in prehistoric doing. It was a conservative but environmentally-friendly approach that relied on intuitive thinking.

As Dr. Ap Dijksterhuis explains in his study "*The beautiful powers of unconscious thought*" (published in Psychological Science Agenda, October 2009):

"*Only among unconscious thinkers was the correlation obtained, indicating that the benefits of expertise become apparent when one thinks unconsciously rather than consciously. Recently, the finding that unconscious thinkers outperform conscious thinkers and immediate decision-makers have been replicated in new and interesting domains.*"

"*Why unconscious thought is helpful is that it seems to be at weighting the relative importance of different attributes. During unconscious thought, important matters become more important, whereas unimportant matters become more unimportant. When we have to make a decision based on numbers and calculations, such as in games, conscious thought usually* (significantly) *outperforms unconscious thought.*"

Other studies on the same subject indicate that <u>unconscious thought is poor at dealing with complexities and complex problem-solving</u>. This appears as the weakest aspect of the Paleolithic mind, where the opticoders were dominant in all mental processes. It also characterized an efficient-conservative mind that was good at repeating the same thing all over again. But here, the adaptability to change was sacrificed.

<u>The A-opticoders were dimensionally reduced to operate linearly, becoming B-opticoders, and some B-opticoders and sound coders adapted to a lexicoding function</u>. It intermediated a direct connection to the linearity of the consciousness. Thus, the lexicoders further propelled a massive conscious development.

As Dr. Andrew Pyzdek explains in one of his articles published in *The World Through Sound* (March 2021):

"*Acoustics is absolutely rife with linearity. Nearly every acoustic system is linear to a high degree. For most sounds that you hear, the air is a linear medium, and the natural vibrations of most objects and resonators are linear, too. Acoustic impedance can give a full description of the behavior of a linear system, using superposition."*

* * *

How do the brain structures operate to assemble the language?

Many studies indicate that dissipative structures can be formed by the periodic oscillation of the forces among the particles and create far-from-equilibrium self-organizing structures.

As Mario Tagliazucchi, Emily A. Weiss, and Igal Szleifer explain in their study (*Dissipative self-assembly of particles interacting through time-oscillatory potentials*, published in PNAS July 8, 2014):

"*Dissipative self-assembly is the emergence of order within a system due to continuous input of energy. This form of nonequilibrium self-organization allows the creation of structures that are inaccessible in equilibrium self-assembly. Self-assembly occurs via the oscillation of interparticle potentials.*"

"*If the input of energy is stopped, dissipative structures are destroyed as the system evolves toward equilibrium; therefore, these structures exist only far from equilibrium; they are unique because they can adapt to environmental changes. Nature excels in using dissipative structures to minimize wasted energy.*"

Energy is continuously injected into the system in the cases described above, making flows of patterns and structures arise.

Quasi-static and dynamic assembly can occur when the frequencies of the components are sufficiently close and/or low, and then the particles follow the beats between the components. It causes incessant restructuring of the particle assemblies. By tuning the gradients, the components can be structured and restructured into a variety of array configurations.

The processes described here have a considerable similarity with how the brain processes language.

According to ScienceDaily from June 19, 2017, (Xiang Zhang et al., 2017):

"Scientists at the Department of Energy's Lawrence Berkeley National Laboratory demonstrated how particles, floating on top of a glycerin-water solution, synchronize in response to acoustic waves blasted from a computer speaker."

"We showed that individually 'dumb' particles could self-organize far from equilibrium by dissipating energy and emerge with a collective trait that is dynamically adaptive to and reflective of their environment."

*"After the researchers intentionally broke up the particle (which were a collection of straws pieces) party, the pieces would reassemble. It was a basis for developing **intelligent networks**."*

The above experiment also shows similarities with the process that develops between the structural elements of a language. As it seems, the sound could break a collection of words and reassemble it as dissipative processes. It gives the collection of words an "intelligence" that did not exist before assembling or reassembling.

According to research in the domain of Dissipative Systems, those mentioned above and many others not mentioned here, the change in the human mental setup, causing language, is related to changes in neural frequencies, which consequently affected the working rhythms.

There is a fair chance that the frequency changes (especially theta band frequencies) became reflected in the mechanism of lexicoders because of their adaptation to low frequencies of the sound processing. The opticoders could not benefit from such low frequencies.

As indicated in their study *(Direct brain recordings reveal hippocampal rhythm underpinning of language processing,* published in PNAS October 4, 2016) Vitoria Piai, Kristopher L. Anderson, Jack J. Lin, Callum Dewar, Josef Parvizi, and Nina Dronkers reveal that *"the hippocampal complex contributes to language in an active fashion, relating incoming words to stored semantic knowledge, a necessary process in the generation of sentence meaning. Theta oscillations, pivotal in memory function, track the amount of contextual linguistic information provided in sentences. Our effects thus reflect the differential role of stronger associations between the words in the sentence to a particular context."*

* * *

The **maneuvering of the language structures** is facilitated by a common covariant character specific to all entities within the linear realm.

Mental assembling is generating verbal collections, which are complexities, and which produce an emergent result. Such an emergent behavior defines the **Intelligence introduced to cognition**.

Why do the lexicoders show physical characteristics that allow high-maneuverability in contrast to features of opticoders?

This essential difference is that sound displays a host of linear properties, while light remains fundamentally nonlinear.

The research indicates that the acoustic processing is intimately related to low frequency rhythms. For memory task with acoustically presented words, Schack & Klimesch (202) found phase coupling networks during encoding in lower alpha band only. Listening to music produces a distinct reactivity also in the lower alpha band. As it seems, the differences between the lower and upper alpha band show various aspects of auditory information processing.

Cognitive domains are associated with frequency ranges of delta 2-4 Hz, theta 4-7 Hz, alpha 8-12 Hz, beta 16-25 Hz, gamma 30-50 Hz. These cognitive areas are mediated by the interference of theta domain that serves the language. The cognitive domain directly associated to language uses delta, alpha and theta frequencies.

It is documented by multiple research that cortical speech tracking in the theta frequency encodes most speech clarity, while delta band encodes the high-level speech comprehension.

The research discovered some emergent persistent localized cortical wave patterns are developed in loop patterns, which appear on brain surface in the intersection areas between excitation (neural spikes) and dissipation. These loops are 1D and 2D and located within 0.5-4 Hz but also in higher ranges as 25-100 HZ. It indicates a clear linear brain wave propagation.

The high frequencies, like beta, are specifically associated to high cognitive processes. But also, upper alpha band indicates increased cognitive activity.

The verbal feeding of cognition offers some much-needed tools, which help achieve better daily problem-solving.

The Science of Complexity states that solving a specific complex problem requires building up a mental complexity at least equal to that of the problem to deal with. Cognition started to build verbal complexities intended to solve daily problems. Lexicoders materialized this new processing.

Current developments seem to drive our brain to resurrect some of those old characteristics. The old mechanism should be noted because it would play an increasing role in the near future.

*   *   *

Every mental complexity provides an emergent result that cannot be exploited further unless stored in the memory. Thus, erasing the old results was counterproductive for the human experience.

Here, the opticoders inspired some exterior storage that took the form of external encoding of symbols and evolved into **symbolic writing**.

The dual-use of visual pathways stimulated and allowed the development of a written language that helped revitalize some old functions of the opticoders.

Today, the opticoders seem to make new inroads by favoring novel visual expressions, like film, TV, and a rapid expansion of imagery tools.

Artificial Intelligence and gaming are the latest advents on such new visual trends of brain processing.

In the US, two-thirds of the population practice computer gaming, and two billion gamers are reported worldwide. The average gaming time is estimated to five to seven hours per day.

As it appears, we face a cycling process, where the **visual tends** to come back. It seems to be at work **a novel adaptation toward more efficient and rapid outcomes vastly offered by visual processing**.

* * *

Let's see the **benefits and secondary effects produced by gaming** because they vastly seem to affect our mental and social future.

Marc Palaus, Elena M. Marron, Raquel Viejo-Sobera, and Diego Redolar-Ripoli, in their study *Neural Basis of Video Gaming: A Systematic Review*, published in Frontiers in Human Neuroscience and online 2017 May 22, explain:

"*Video gaming is an increasingly popular activity in contemporary society, especially among young people, and video games are increasing in popularity not only as a research tool but also as a field of study."*

"It has been possible to establish a series of links between the neural and cognitive aspects, particularly regarding attention, cognitive control, visuospatial skills, cognitive workload, and reward processing."

"An important segment of society, over 30% in tablet computers and 70% in smartphones, has been exposed to these technologies and can be considered now, in some form, causal gamers."

"There are articles which study the effects of VG exposure over the nervous system and over cognition.; therefore, we should be able to appreciate changes in the neural bases. They evaluate the possible transfer effects of VG training to wider cognition domains."

"Brain activation patterns depend on the cognitive demands of the environment and also on the associated level of workload (Vogan et al., 2016), which is directly related to the allocation of resources to the working memory and its attentional processes (Barrouillet et al., 2007)."

"We see the result of neural recruitment mechanisms as the cognitive demands increase (Bavelier et al., 2012a)."

"It is likely that functional changes related to the manipulation of cognitive load appear along with the attentional networks and in specific key nodes related to executive functions, mainly in prefrontal and parietal cortices."

"The theta band power band increased. The theta band power also displayed higher power compared to the resting condition and gradually increased during the gameplay. (Sheikholeslami et al., 2007). The theta band seems to be directly related to the level of cognitive demand in a wide variety of cognitive abilities, such as attention, memory, and visuospatial processes. On the other hand, beta band

power seems to be more associated with the complexity of the task, especially in frontal and central areas, likely indicating a qualitative change in the cognitive strategy followed by the participant or the type of processing done by the brain (Brookings et al., 1996)."

"Frontoparietal activity, linked to the attentional process, also exhibits recruitment effects as game difficulty increases. Increased blood flow in prefrontal areas like dlPFC was also associated with increased cognitive demands related to attention, verbal and spatial working memory and decision making (Izzetoglu et al., 2004)."

"Most changes regarding cognitive control observed after VG plays like be detected in parietal areas and basal ganglia. Prefrontal regions are one of the brain areas in which GM volumetric changes have been observed as a result of cognitive training with a VG, which is remarkable if we consider that the common VG training period spans from a few weeks to a couple of months. These volumetric changes even result in correlation with transfer effects in cognitive control tasks (Hyun et al., 2013). Volumetric-behavioral correlations work both ways since individuals with decreased orbifrontal cortex (OFC) volumes due to VG addiction show poorer performance in similar tasks (Yuan et al., 2013). Still, prefrontal activity is not only affected by the complexity of the task, but also by the nature of the task and the individual differences between participants (Biswal et al., 2010)."

"Changes in functional activity after a training period in other executive-related nodes, such as superior parietal lobe (SPL), have also been associated with working memory improvements (Nikolaidis et al., 2014). Connectivity-wise (Martinez et al., 2013) found rest-state functional connectivity changes in widespread regions (frontal, parietal, and temporal areas) as a result of the VG training program, which was attributed to the interaction of cognitive control and memory retrieval and encoding."

"By studying lifelong experts or professional gamers, some studies have detected structural GM changes that correlated with improved executive performance, involved posterior parietal (Tanaka et al., 2013) and frontoparietal (Hyun et al., 2013) regions. Regarding structural connectivity, WM integrity changes in thalamic areas correlate with improved working memory, but the integrity of occipitotemporal fibers had the opposite effect (Strenziok et al., 2014). VG experience also seems to consolidate the connectivity between executive regions and the salient network and responsible for bottom-up attentional processes. Training older adults in a strategy VG seems to improve verbal memory span (Mc Garry et al., 2013), but not problem-solving or working memory, while using a 2D action VG improved everyday problem solving and reasoning."

The study conclusions are:

-attentional benefits resulting from the use of VG seem to be the most evidence-supported aspect;

-optimization of cognitive costs in visuomotor tasks performance is commonly observed;

-some regions show volumetric increases that are thought to be directly related to visuospatial and navigational skills;

-decreased activation in the associated regions;

-VG training paradigms showed improvements in cognitive control related functions, particularly working memory;

-importance of frontoparietal activity for VG purpose;

-the role of the reward system is also present.

The study found several adverse effects:

-reduced recruitment in the ACC and possible reduced attentional skills;

-adverse effects on social information processing, leading to individual isolation;

-delayed microstructure development in extensive brain regions;

-lower verbal IQ.

Specifically, the right hippocampus, right frontal cortex, and cerebellum showed a significant increase in volume due to many years of gaming (professional gamers).

During MRI procedures on professional gamers, it was observed an increase in neuroplasticity and neurogenesis. The gamer's gray matter gains were observed exclusively in the right hemisphere, known to host visual abilities. They can be related to an undetected yet stimulus developing in the neurogenesis processes. It could be a path neurogenesis took before language that now resurfaced during VG's very long practice (of at least ten years or more).

In my opinion, all beneficial outcomes are strictly related to the combined use of the white matter pathways of the opticoders and lexicoders. The opticoders take advantage of this visual window of opportunity and reinforce the old visual attention processing.

However, as this large reviewing study enhances but not discloses, during the VG practice, a functional and working transfer seems to occur between opticoders and lexicoders. This transfer results in stimulating the existing cognition. It is known that cognition was significantly developed and restructured by the activity of lexicoders.

An aspect not reached by the mentioned study refers to the brain energy cost of gaming. It is known that visual activity costs the brain a significantly larger amount of energy versus the adopted verbal mode.

The study mentions various volumetric gains occurring in the right hemisphere due to VG. This is in line with a traditional visual activity that employs more power and results in higher energy consumption.

Maybe at present, due to the combined work of opticoders and lexicoders, the energy consumption does not sensibly increase because of these coders' sharing. Even then, many studies report a significant energetic exhaust (tiredness) of the gamer. As it seems, an appropriately combined physical exercise could supplement additional energy to the gaming brain while increasing gaming efficiency.

As various studies report, the gaming activity increased from 5.1 hours in 2011 to over 7.2 hours in 2020.

* * *

Another important aspect of gaming is its **influence on psychological functioning**.

From a study authored by Juliane M. von der Heiden, Beate Braun, Kai W. Muller, and Boris Egloff (*The Association Between Video Gaming and Psychological Functioning*), published in Front. Psychol., 26 July 2019, I like to quote some of their findings.

"*Drawing on a large sample, our results revealed a medium-size relation between potentially problematic video game use and poor psychological functioning with regard to general psychological symptoms, maladaptive coping strategies, negative affectivity, low self-esteem, and a preference for solitude as well as poor school performance. These findings are in line with those of prior work (e.g.., Kuss and Griffiths, 2012; Milani et al., 2018).*"

"*Specifically, distraction-motivated gaming went along with higher symptom rating, lower self-esteem, and more negative affectivity, whereas playing to establish social relationships in the virtual world was related to a large number of online connections and more positive affect while playing.*"

"*Finally, positive effects while playing and a large number of online friends were the strongest unique predictors of potentially problematic video game use, followed by psychological symptoms, a lack of offline connections, and poor school performance. Poor psychological functioning seems to be a unique risk factor for potentially problematic video gaming."*

The study found:

-games allow people to avoid everyday problems and instead immerse themselves in another environment;

-the games reduce the amount of time and the number of opportunities gamers have to practice real-life behavior;

-players use escape-oriented reasons such as distraction;

-playing for gain-oriented reasons such as storyline or the social connections in the game is the only positive psychological outcome;

-but this same positive outcome may reveal a hidden addiction toward general gambling habits;

-gaming is associated with obesity in teens due to alleged food intakes during gaming as an activation of the reward mental centers;

-there are gaming-associated vision problems, headaches, poor concentration, and even seizure occurrences.

A study by Greg L. West, Brandi Lee Drisdelle, Kyoko Konishi, Jonathan Jackson, Pierre Jolicoeur, and Veronique D. Bohbot (*Habitual action video game playing is associated with caudate nucleus-dependent navigational strategies*) published by Proceedings of the Royal Society B, May 2015 has concluded:

"However, we also found that gamers rely on the caudate-nucleus to a greater degree than non-gamers. Past research has shown that people who rely on caudate-nucleus-dependent strategies have lower gray matter and functional brain activity in the hippocampus. This means that people who spend a lot of time playing video games may have

reduced hippocampal integrity, which is associated with an increased risk of neurological disorders such as Alzheimer's disease."

* * *

However, **thinking hard uses calories, but the energy use is minimal**. It is not enough to burn any of the body fat. Our brain burns more energy at rest compared to running a distance. The brain energy comes from glucose that enters the bloodstream while partially ending its route into the brain.

Regarding the energetic needs of the brain, I like to quote Dr. Doug Boyer, an associate professor of evolutionary anthropology from Duke University that published a study along with the researchers Harrington and Messier (November 2019, published in Live Science).

The authors explain:

"The brain can shunt blood (and thus energy) to particular regions that are being active at that point. But the overall energy available in the brain remains constant. So, while there might be significant spikes in energy use at localized regions of the brain when we perform difficult cognitive tasks when it comes to the whole brain's energy budget overall, these activities don't significantly alter it."

* * *

Before the language's new abilities, the Paleolithic brain dealt with visual mental possibilities expressed unconsciously.

I would say it was an "intuitive" solution because it came from the unconscious. And, it was suitable for most of the time when pattern recognition occurred.

However, a sudden change in the environmental conditions implies a new pattern that does not match the pattern recognition of the previous memory.

Humans were confined to African environments for very long. They accidentally traveled out of a few fertile African environments; these people could not reach too far away and settle before something was producing a cause that perished them.

It was no sufficient mental adaptability and plasticity that would allow them to mount a cognitive complexity at least the size of the problem to solve. At that time, these people could not mount complex thoughts.

So, before the language, the Paleolithic visual brain cannot generate complexity in response to the natural intricacies of the environment.

It clearly indicates that visual thinking, hearing, olfaction were not able to produce mental complexities.

Through language development, humans acquired a powerful tool to assemble things in response to challenges by stimulating a **mental dissipation**. Consequently, such a tool made them able to compete with most challenges.

But, in today's world, we encounter a new tendency toward returning to visual thinking.

*   *   *

I like to conclude by posing several questions about our future evolution.

The **first question**: if the visual brain cannot assemble mental complexities to deal with the complexity inserted in the VG, what other feature does it?

The studies quoted above indicate that the opticoders work combined with lexicoders. Hence, the lexicoders still assemble those needed mental complexities. Or, the opticoders would turn to exploit a transferred behavior from lexicoders.

The **second question**: if, in the long term, gaming stimulates the right hemisphere functions (as in professional gamers), would these functions gradually annihilate the verbal mode?

We hope that visual and verbal will continue to work together in a somehow strange evolutionary way.

The **third question**: if the visual mode will continue to produce neuronal gains (gray matter) into the right hemisphere, would this cause brain enlargements?

However, we know that previous augmentation or diminishing of brain volume needed many thousands of years to be accomplished. Thus, we cannot expect any sharp change in our brain volume.

The **fourth question**: if the brain would not enlarge, what will happen with the excess gray matter steadily generated in the right hemisphere of all long-term gamers?

It might cause a severe anatomical problem in the future gamer's brain.

The **fifth question**: if the energetic consumption of the brain will continue to increase, how will the brain solve this problem in a short interval?

Maybe the physical exercise will not provide enough energy to compensate for the advancing brain consumption. Other anatomical and functional changes would probably occur, like a new reduction in the speed of processing.

The **sixth question**: would changes in the gamer brain allow mental assembling at the same rate as it currently does?

Here, we may expect a negative shift in such assembling activity that would significantly reduce the "intelligent" outcome.

The **seventh question**: would generalized gaming severely increase the social and psychological problems mentioned early?

At present, this seems to be the case. If the worldwide authorities did not intervene sharply to regulate the nature and content of the games, things would escalate in the wrong direction. At the same time, many generations will become fully compromised.

A chronic metabolic oxygen deficiency

Abstract

The HBOT (hyperbaric oxygen treatment) research that I investigated indicates such therapy's success for various diseases. The studies suggest that the HBOT could produce partial or total remission for particular illnesses in each experimental medicine field.

I tried to evaluate the global significance of our civilization. It was clear that hyperbaric oxygen has a role in supplementing our organism with an amount of needed oxygen. Why was this supplemental oxygen so necessary and beneficial?

I concluded that our human body experiences a chronic oxygen-deficiency that increased exponentially in recent times.

I like to mention that The Nobel Prize in Physiology or Medicine in 2019 was awarded to G. Kaelin Jr., Sir Peter J. Ratcliffe, and Gregg L. Semenza to discover how cells' sense adapts to oxygen availability. Hence, their research is fundamental in understanding the exact biological mechanisms that regulate oxygenation processes. I am not in the capacity to comment or add their work.

Introduction

During my investigation, I was able to identify a common cause for such oxygen deficiency: it refers to our *lifestyle*.

I can compartment it into four distinct aspects, which are interrelated:

-the *indoor lifestyle* of the last five generations or so is the most important outcome for this oxygen deficiency; in developed countries, there is a 90%-93% indoor life, while the developing world is catching the gap at high rates;

-the *food intake* was radically changed due to agricultural practices; after the Industrial Revolution, it changed again in a more effective manner due to artificial fertilizers, insecticides, MGOs practices, and novel food processing and packaging;

-the substantial *reduction of physical activities* as a result of various types of machinery; such use of various machinery gradually replaced human and animal energy with mechanical energy activated by fossil fuels;

-*social activity-related stress* becomes a major detrimental trend responsible for inflammation, hypoxia, and accompanying diseases.

All these four aspects, characterizing our modern lifestyle, incrementally and cumulatively contribute to a significant reduction in our body systems' oxygen production.

Today's human experiences a *chronic hypoxia phenomenon* primarily unnoticed by the regular research, but it surfaces when the HBOT is applied. Such Hypoxia seems to cause the majority of our modern diseases and illnesses.

In the meantime, medical literature documents the case of hyperoxia, where an excess of oxygen, resulting from biological processes, is not eliminated by the process of reoxygenation. So, many toxins and radicals produced during hyperoxia are improperly annihilated by a weak reoxygenation process.

It also shows a lack of sufficient oxygen and places where the human body needs supplemental oxygen. Such unprocessed waste also contributes to disease. So, both hypoxia and hyperoxia would equally evolve into a medical condition because the organism runs short of necessary oxygen resources for accomplishing these jobs.

What would unveil this reality? The answer is the hyperbaric oxygen treatment (HBOT) can equally fix, at least in part, both hypoxia and hyperoxia.

I would say all diseases can be treated, at least in part, by hyperbaric oxygen, manifest a noticed or an unnoticed hypoxia and hyperoxia. Both processes show an oxygen deficiency that causes disease.

In our time, the mentioned phenomenon of oxygen-deficiency rapidly accelerates. The most notorious indication of this alarming situation comes from a recent *dramatic increase in sensitivity to viruses* that currently troubles us with the Covid-19 pandemic manifestation.

Such a pandemic imposes human isolation by areal lockdowns, social distancing, and many other severe measures, ultimately affecting the economies, interrupting international trade and tourism, and introducing significant stress among the affected populations.

The *virus of this pandemic is prone to influence for long-term many other systems within the body*.

There is little information yet, but it seems that it may exist the possibility that the human virus, in unknown conditions, can migrate into animals that could devastate them in significant proportion.

These threats are real and quite severe, while they may add more harm to our civilization, exacerbating all other outcomes of the crisis of oxygen-deficiency.

I briefly like to discuss the consequences of oxygen deficiency:

-the first outcome is an *accelerating spread of disease and/or illness* because the same oxygen-deficiency process results in each occurrence of the disease;

- a novel development occurs in the last 60-70 years and was significantly accelerated in the previous 20 years; it is a process of increased human susceptibility or *high sensitivity to viruses*;

-another detrimental outcome is a *decrease in neurogenesis* that already had and would continue to diminish cognition and brain plasticity for a significant number of generations; in the media, this process is popularly called ideocracy.

Developments

Hypercapnia (as vasodilation) can occur when one breathes high carbon dioxide. The lungs entirely expel excess carbon dioxide. But here, the process of hypercapnia makes the body produce 36%

higher levels of nitric oxide that acts on tissues and membranes as a vasodilator. The result is that the body systems that experience hypercapnia can lose a part of their needed intracellular oxygen.

Thus, a high concentration of carbon dioxide in the ambient causes oxygen deficiency in different systems and indirectly induces high nitric oxide levels.

During *hypocapnia* (vasoconstriction), the nitric oxide level decreases by 30%. There is known that hyperbaric oxygen produces vasoconstriction. It diminishes certain nitric oxide levels, helping the systems and organs benefit from a better oxygen balance.

Practices and results of HBO related procedures

Jae W. Lee, Junsuk Ko, Cynthia Ju & Holger K. Eltzschig, in their paper *Hypoxia signaling in human diseases and therapeutic targets* (published in Experimental & Molecular Medicine 51, 1-13 -2019), explain:

"Over 95% of the oxygen that is delivered into the capillary vessels through the alveolar-capillary exchange system binds to hemoglobin. The heart pumps oxygenated blood to the periphery, which is crucial for organs and cells to function and perform oxidative phosphorylation. Hypoxia can result from any failure that might happen during this process, insufficient blood flow to an end organ, dysfunctional or low hemoglobin levels, or chemically induced Hypoxia. Hypoxia activates the hypoxia signaling pathways, which is predominantly governed by hypoxia-inducible factor (HIF) stabilization".

"Under hypoxia, the activity of PHDs and FIHs is suppressed. All of these studies support the notion that hypoxia and inflammation have an interdependent relationship. Many studies demonstrate that although hypoxia can cause tissue inflammation,".

Debora Coimbra-Costa of the University of Barcelona, Spain, in a paper published in Redox Biology on February 24, 2017, explains:

"There is a constant balancing act between the creation of reactive oxygen species (ROS) and their elimination by the antioxidant system. Abnormally low levels of ROS can impede cellular signaling and some normal intracellular reactions". On the opposite, "high levels of ROS create oxidative stress that leads to improper oxidation of lipids, proteins, and DNA, causing cellular and neuronal damages."

Coimbra-Costa's study determined that "oxidative stress could be considered the main damaging event induced by hypoxia."

However, "Reoxygenation after hypoxia does not induce further oxidative damages. Increased blood flow promotes the resupply of oxygen and removes harmful waste products, such as lactate and hydrogen, that otherwise may cause cell and neuron damages. Apoptosis after hypoxia was more prevalent in the hippocampus, while it affected cognition and brain plasticity".

Alfredo J. Garcia III, Robert W. Putman, and Jay B. Dean, in their paper *Hyperbaric hyperoxia and normobaric reoxygenation increase excitability in CA1 hippocampal neurons* published in the Journal of Applied Physiology (September 2010), state:

"Reoxygenation (from 0.0 or 0.6 ATA O2) usually produces a response similar to that of HBOT. Whether caused by breathing HBO or Reoxygenation following hypoxia, acute hyperoxia stimulus is a powerful stimulant for orthodromic activity and neural plasticity in CA1 hippocampus".

My note: hyperoxia is the opposite of hypoxia; it refers to an excessive resupply of oxygen. (The supplementation of oxygen can lead to oxygen toxicity that can affect the brain, pulmonary system, and the ocular mechanism).

Ryan Choudhury in *Hypoxia and hyperbaric oxygen therapy: a review* (published by International Journal of General Medicine in November 2018) says:

"In the US, diseases such as stroke, cancer, heart disease, and chronic lung disease-account for a majority (up to 60%) of the total number of deaths. Hypoxia is a significant component of the pathology of these conditions as the unavailability of oxygen leads to physiological responses that, if not resolved, progress to localized hypoxic responses, cell metabolic inefficiency, organ dysfunction, and final death".

It is known that after migration to a particular site of infection, neutrophils engage the exposed invaded organ with antimicrobial peptides and proteins. Still, here the most efficient product by far is the reactive oxygen species.

Also, it is known that impaired neutrophil function predisposes to infection with pathogens. Here, the infiltrating immune cells and replicating bacteria consume oxygen, producing profound tissue hypoxia. Hence, almost all sites of infection are characterized by tissue hypoxia.

Katheryn Begg and Mahvash Tavassoli, in their paper *Inside the hypoxic tumor: reprogramming of the DDR and radioresistance* (published on Cell Death Discovery on April 2020), explain:

"The hypoxic tumor is a chaotic landscape of struggle and adoption. Against the adversity of oxygen starvation, hypoxic cancer cells initiate reprogramming of transcriptional activities, allowing for survival, metastasis, and treatment failure. This makes hypoxia a crucial feature of aggressive tumors".

"Hypoxia is present in every solid tumor, a certainty of cancer's characteristics disorganized and functionally inefficient vasculature, rapid growth, and demanding metabolism. The result is a comprehensive re-writing of transcriptional programs.

After decades of research, it has become clear that the relevance of hypoxia for both oncogenesis and treatment resistance is inescapable".

Hyperbaric oxygen slowed the growth of some cancers but not others. However, other scientists consider that higher than average oxygen concentration has not been proved scientifically to kill the disease.

Paul Eggleton, senior lecturer in Immunology, University of Exeter (UK), in his paper *Multiple sclerosis survivors swear by hyperbaric oxygen-but, does it work?* Stated:

"In the MS case, people self-prescribe hyperbaric oxygen, which is delivered to them by trained operators. The idea to use oxygen as a treatment for MS began over 45 years ago, in 1970, when two Romanian doctors, Boscetty and Vernoch, treated patients with brain injuries with prescribed oxygen".

"The clinical, regulatory bodies in the US and the UK do not feel the clinical trial evidence is strong enough to endorse the procedure, yet thousands of people in the UK and elsewhere continue to treat themselves with hyperbaric oxygen. Between 1982 and 2011, over 20,000 people with MS in the UK used hyperbaric oxygen over 2.5 million times. Many of 100,000 MS in the UK have taken charge of managing their treatment with hyperbaric oxygen".

"The brain's ability to repair some of MS damages helps people to feel better for a while before relapsing once more. When the disease becomes chronic, the ability to damage and undergo remission declines".

In another paper by Hilary D. Wilson, Virginia E. Toepfer, Arun K. Senapati, Judy R. Wilson, and Perry N. Fuchs (published in Elsevier on August 9, 2007) indicate that hyperbaric oxygen therapy significantly reduced joint inflammation in a study on animals' hyperalgesia.

Many other studies in the field mentioned above also indicate that hyperbaric oxygenation for arthritis or osteoarthritis treatment is very efficient because the increased arrival of oxygen to joint and tendon tissue rapidly reduces inflammation and operates repairs.

The Congress on Hyperbaric medicine indicates the overall practical results in the combined fields:

-remission 23.4%;

-obvious effect 51.4%;

-improvements 16.2%

-no effect, 8.1%.

It is known that oxygen helps bring about specific chemical reactions that result in energy production. An oxygen-deprived body develops tissue hypoxia and chemical-induced hypoxia, which relate to a significant reduction in all body functions. Many diseases impair oxygen utilization.

The case of atrial fibrillation (AF) treatment with HBOT has not been reported yet, in the literature. Even then, several cases have been documented by specialists indicating that after HBO therapy, ECG showed that the rhythm had returned to normal sinus rhythm when an amiodarone treatment was supplemented.

A group of scientists from Louisiana State University Health Science Center lead by Dr. Paul Harch and Dr. Edward Fogarty reported on January 24, 2019, a clinical case of Alzheimer disease, where a patient, after HBO treatments-five day a week for 66 days, showed a global 7-38% improvement in brain metabolism that is the most considerable improvement of any therapy known for Alzheimer disease.

Ronit Shapira, Shai Efrati, and Uri Ashery from Tel Aviv University published the paper *Hyperbaric oxygen therapy as a new treatment approach for Alzheimer's disease* (published in Neural Regeneration Research, May 18, 2018).

The authors indicate that "AD, the most common form of dementia in the elderly, accounts for 60-80% of all dementia cases". "HBOT has been shown to improve neurological functions and life quality following neurological incidents such as stroke and traumatic brain injury, and to improve the performance of healthy subjects in multitasking."

"A recent study was demonstrating that HBOT can ameliorate AD-related pathologies in an AD mouse model. A triple-transgenic mouse model was exposed to 14 days of HBOT and showed reduced hypoxia and neuroinflammation, reduction of beta-amyloid plaque and phosphorylated tau, and improvement in behavioral tasks."

The authors suggest that "oxygen is an important tool in the arsenal for the fight against AD. This approach presents a new platform for the treatment of AD (Alzheimer Disease)."

"Studies have shown that in elderly healthy subjects, oxygen supplementation improves the subjects' performance in cognitive tasks and changes the electroencephalographic (EEG) pattern of the brain activity, indicating that oxygen is a rate-limiting factor in normal and disease-associated cognitive function."

They concluded that HBOT significantly reduced the presence of hypoxia in the hippocampal formation.

Neuropsychological changes, as a result of systemic reduction of oxygen under hypoxic conditions, cause various impairments, like the alteration of mood, behavior, fatigue. Still, most of all, it affects cognitive function. This is so because it is demonstrated that hypoxia is linked to changes in brain activity.

Hypoxia develops as a result of physical inactivity. By contrast, exercise is well known to improve cognition, mood and eliminate fatigue and depression.

Hypoxia could alter the decision-making process by ignoring the obvious disclosed by cause-effect interaction, altering regular judgment toward depression, tension, and confusion.

In the end, hypoxia is seen as a significant cause that may affect our capacity to survive in edging conditions.

HBOT has been reported to improve mitochondrial function and alter the balance between glycolysis and mitochondrial respiration, countering the effect of viral infection in cellular caloristasis networks and improving hypoxia Covid-19 patients.

The hyperbaric oxygen treatment adds a substantial supply of oxygen into the bloodstream, directly treating the hypoxemia.

Hyperbaric Medical Solutions listed on their site several papers that refer to HBOT applications for Covid-19 patients.

Many studies have indicated that HBOT applications improve kidney function after infection and reduces kidney damage in diabetic patients.

HBOT is an application that significantly diminishes UV skin damages.

Qiangwei Fu, Sean P. Colgan, and Carl Simon Shelley published a paper, *Hypoxia: The Force that Drives Chronic Kidney Diseases*, published in Clinical Medicine and Research in March 2016.

The authors stated:

"In the US, the prevalence of end-stage renal diseases research epidemic proportions in 2012 with over 600,000 patients being treated. The rates among the elderly are disproportionally higher. The number of patients treated for terminal kidney failure worldwide has continued to grow at an annual rate of 7% that stays against the annual growth of the global population of 1%".

"It was found that this endothelium-independent mechanism involves leukocytes directly sensing hypoxia and responding by transcriptional induction of genes. This induction maintains the long-term inflammation by which hypoxia drives the pathogenesis of CKD".

Qi-Zhoung Lu, Xiang Li, Jun Quyang, Ji-Quan Li, and Gang Chen published the paper *"Further application of hyperbaric oxygen in prostate cancer"* (published in Medical Gas Research in December 2018).

The authors indicate that they used HBOT as an adjuvant treatment for the effect of oxygen on carcinoma, decreasing medical complications. The authors explain that when the cells get enough oxygen in the microenvironment, the surrounding cells become active and replicate effectively, preventing the spread of cancer.

Neural hypoxia-discussion.

Maggie A. Khuu, Chelsea M. Pagan, Thara Nallamothu, Robert F. Hevner, Rebecca D. Hodge, Jan-Marini Ramirez, and Alfredo J. Garcia III published a paper titled *Intermittent Hypoxia Disrupts Adult Neurogenesis and Synaptic Plasticity in the Dentate Gyrus* (Journal of Neuroscience, February 13, 2019).

Authors show that "individuals with sleep apnea often exhibit changes in cognitive behavior and alterations in the hippocampus. Intermittent hypoxia (IH) perturbs multiple aspects of adult neurogenesis, increases the proportion of radial glial cells in the subgranular zone, yet decreasing the proportion of adult-born neurons in the dentate gyrus".

The authors indicate that the HBOT prevents adult-born neuron loss.

IH is found to cause "both reactive oxygen species-dependent and reactive oxygen species-independent effects on adult neurogenesis and synaptic plasticity in the dentate gyrus."

"Adult neurogenesis uniquely supports the dentate gyrus by providing a source for cellular heterogeneity among the principal cells of this network. When compared to relatively older and more mature counterparts, new adult-born granule cells are more excitable. New neurons enhance synaptic plasticity for a limited time. Thus, conditions that alter hippocampal adult neurogenesis are likely to impact hippocampal neurophysiology as well".

In conclusion, this study suggests that "the impairment of synaptic plasticity was accompanied by reduced adult neurogenesis." "The sleep apnea (because of its hypoxic nature) may be a condition that dictates the outcome of hippocampal adult neurogenesis and synaptic plasticity" that "contribute to a decline in neurocognitive behaviors."

Discussion of the effect of increased carbon dioxide levels on neural functions

I researched several studies investigating the influence of carbon dioxide on brain activity and which influences its metabolism.

All recent studies found that increased carbon dioxide levels cause the brain to reduce metabolism and spontaneous neural activity while the brain enters a low arousal state. It was found a selective suppression effect on resting-state neural activity. Dynamic measurements of brain metabolism revealed that a manifestation of mild hypoxia induced by the carbon dioxide to the brain produces a suppression of cerebral metabolic rate of oxygen by 13.4 to 16.8%.

Discussion on viral sensitivity

The carbon dioxide, which is a vasodilator, ultimately generates, in the long-term, a decline in neurocognitive behaviors.

The indoor lifestyle (estimated to amount to as much as 90-93% of our lifetime in the developed countries), along with other elements associated with our lifestyles, like physical activity and diet, all contribute to significant processes of hypoxia, which are responsible for most of the modern diseases. Still, they inflict severe damage to our brain, from mental disorders to a steady decline in our neurocognitive capacity (that is what I called earlier in this text under the name of ideocracy).

I like to mention that the literature contains many papers that define stress as causing vasodilatation that favors inflammation and illness. As many scientists consider, stress is the leading cause of many diseases or conditions.

Unfortunately, stress became a significant component of our modern lifestyle, while its role defines another oxygen-deprivation process in the human body.

However, for the last three generations (or the previous 60 years), our lifestyle seems to have crossed a natural balance, at least in the developed countries.

The mentioned above situation is revealed by the increased indoor living containment that diminishes any physical activity. Changes in agriculture (especially the introduction of MGOs) contributed to further changes in our diet.

This broken balance is supported by a *sudden sensitivity to viruses* that found our civilization economically and administratively unprepared and with little knowledge to respond to it.

Several studies on Covid-19 indicate that silent hypoxia is the first element that manifests during this initial phase of this virotic disease.

The sickness evolves from hypoxia's shallow levels toward levels where the bloodstream's oxygen diminished as low as 40-50%.

In a few months since this pandemic began to affect our society, it turns evident that the *stress* being caused by the potential dangers posed by this viral infection has a very negative effect on the general population and seems to favor the catching of this disease. Such stress is another producer of hypoxia, and it can accelerate the evolution of the silent hypoxia developed by the viral infection.

However, prolonged hypoxia tends to strain the heart and almost all other bodily systems. The survivors may expect multiple long-term manifestations of the damages induced by the Covid-19 type of hypoxia to our body's organs and systems.

A new hypothesis on Covid-19: a multisystem vasculopathy

The Journal of Thrombosis and Thrombolysis published on July 5, 2020, a paper with the title *COVID-19: Are we dealing with a multisystem vasculopathy in the disguise of a viral infection?* The article was written by a group of researchers lead by Ritwick Mondal, Durjov Lahiri, and Shramana Deb.

The authors indicate that:

"Apart from the primary respiratory symptoms like fever, breathlessness, and fatal pneumonia, a subset of patients presenting with derangement in vascular parameters have also been documented through clinical and pre-clinical reports."

"Based on the existing case reports, literature and open databases, we also analyze the differential pattern of vasculopathy related changes in COVID-19 positive patients".

The authors support the hypothesis that vascular endothelium is a crucial target of COVID-19. It "manifests as a systemic vascular disease with extrapulmonary dissemination among various organ systems. It is evidential in the form of viral shedding in respiratory secretions, stool, urine, and even sweat".

"Patient reports suggest that MERS-CoV infection (that shares 50% phylogenic homology to SARS-CoV) targets multiple systems including pulmonary, cardiovascular, renal, coagulation, gastrointestinal tract, muscular". These similarities led the research group to found an enzyme (ACE-2) responsible for such conversions and multisystem dissipation.

However, the epidemiology of COVID-19 has been significantly more extensive than in the case of MERV-CoV, and therefore, the number of affected individuals increases exponentially. In both cases, the indication that we deal with systemic vascular disease becomes pertinent. Even it can be concluded that, in the case of viral infections, like COVID-19 and MERV, we deal with an acute systemic vascular hypoxia that may have a base in chronic systemic hypoxia due to our contemporary lifestyle.

Covid-19 appears as the first disease on record that tends to affect the global economy as a whole, severely disrupting many types of businesses while changing our basic social behaviors and practices. It may ultimately change the foundation of our lifestyle.

Conclusion

Never our civilization was confronted with so many grave and disruptive challenges (from global warming to a level of pollution that endangers the survivability of all other species to Artificial Intelligence alleged threats to global pauperization and a current global pandemic with no end in sight).

It is evident that, unknowingly but carelessly, our Anthropocene development had disrupted an unknown fundamental balance, bringing ourselves into a crisis that I have called the *oxygen-deficiency crises*.

If such a crisis continues to accelerate its development, it may incur an existential threat.

Acknowledgment

This paper is not intended to enlighten the merits or deficiencies of HBOT technology. It uses hyperbaric oxygen scientifically proven outcomes to validate a chronic oxygen deficiency concept that, in my opinion, it is something that our civilization currently experiences.

5th International Conference on New Findings on Humanities and Social Sciences

Certificate of Virtual Presentation

Presented to

Dan M. Mrejeru

Who participated in the 5th International Conference on New Findings on Humanities and Social Sciences and presented the paper

"Our civilization experiences an existential threat caused by a chronic oxygen deficiency"

Farzam Chakherlouy
Chairman of Organizing Committee

20 - 22 November, 2020
Paris - France
www.hsconf.org
HSCONF-7060

My motivation

I worked all my life as an Engineer Geologist while I kept my scientific background open to a latter eventuality of research and writing on scientific themes of my choice.
I decided to write like an exercise in English around 2003.
My choice for science writing was backup by personal reasons:
-my English grammar was poor because I did not attend any schooling in the English language;
-I was familiar with science reading and its international terms;
-as a scientist, I developed a professional curiosity about how and why things are different from how they appear to us;
-in the meantime, I was a skeptic about some generally expressed ideas that I knew have second motifs;
-the most supporting was the Geophysical training that allowed me to think about a possible correlation between geophysical events and our brain development; it was left to find the biological connection that services the two. My new training in the Science of Complexity played a fundamental role.
The aim of writing this book was to demonstrate, while supported by credible arguments, that the correlation between the geophysical events, processes of complexification, and the human brain's biology could be reasonable explained.

Printed by Libri Plureos GmbH in Hamburg, Germany